How Cancer Works

Lauren Sompayrac, Ph.D.
Retired Professor
Department of Molecular, Cellular, and Developmental Biology
University of Colorado
Boulder, Colorado

JONES AND BARTLETT PUBLISHERS
Sudbury, Massachusetts
BOSTON TORONTO LONDON SINGAPORE

World Headquarters
Jones and Bartlett Publishers
40 Tall Pine Drive
Sudbury, MA 01776
978-443-5000
info@jbpub.com
www.jbpub.com

Jones and Bartlett Publishers Canada
2406 Nikanna Road
Mississauga, ON L5C 2W6
CANADA

Jones and Bartlett Publishers International
Barb House, Barb Mews
London W6 7PA
UK

Production Credits
Executive Publisher: Christopher Davis
Production Manager: Amy Rose
Associate Production Editor: Renée Sekerak
Editoral Assistant: Kathy Richardson
Marketing Associate: Matthew Payne
Manufacturing Buyer: Therese Bräuer
Composition and Art Creation: Dartmouth Publishing
Cover Design: Kristin E. Ohlin
Printing and Binding: DB Hess
Cover Printing: DB Hess

Cover Photo Credit: © Science Photo Library / Photo Researchers, Inc.

Library of Congress Cataloging-in-Publication Data

Sompayrac, Lauren.
How cancer works / Lauren Sompayrac.— 1st ed.
p. ; cm.
Includes index.
ISBN 0-7637-1821-1
1. Cancer—Popular works.
[DNLM: 1. Neoplasms—pathology. QZ 200 S697h 2004] I. Title.
RC263.S645 2004
616.99'4—dc22

2003017752

Printed in the United States of America
07 06 05 04 03 10 9 8 7 6 5 4 3 2 1

DEDICATION

I dedicate this book to my sweetheart, my best friend, and my wife: Vicki Sompayrac.

Contents

Acknowledgments

I especially want to thank two friends who reviewed the entire manuscript. First, my old friend, Dr. Bob Mehler (who is a "real" doctor), paid special attention to the more clinical aspects of this book, and saved me from making a number of mistakes in that area. Without you as my reader, Bob, I never would have attempted this book.

Second, I want to thank a new friend, Professor Albey Reiner. Over the past decade and a half, Albey has taught an undergraduate cancer course to more than 10,000 non-science majors at the University of Massachusetts at Amherst. As you can imagine, Albey has an excellent feel for the features of a book on cancer that would appeal to individuals who have little or no background in science—the target audience for my book. His comments and suggestions have been invaluable in helping me match the level of my writing to the background of my readers and in eliminating topics that seem interesting to me, but which would not hold the attention of a non-scientist. Thank you, Albey. I really feel like we wrote this book together.

I would also like to thank Dr. Charles Wilcox from the University of Alabama at Birmingham for his generous contribution of the photograph used in Chapter 6, and Vicki Sompayrac, whose wise suggestions helped make this book more readable, and whose copy editing was invaluable in preparing the final manuscript.

Finally, I would like to thank "my" editor at Jones and Bartlett, Chris Davis. This book was written at Chris' suggestion, and, as always, it has been a joy to work with him on this project.

About This Book

There are books which discuss cancer from a historical perspective, tracing this disease through the ages. These books tell interesting stories, but they won't teach you much about how cancer works. There are also books that will tell you just about everything that is known about every cancer, but most readers don't need (or want!) this level of detail. Even the best cancer books tend to be "books of lists." While it may be informative to read the first few entries in these lists, it quickly gets boring—and confusing. In fact, some of these lists are so long that even the person who wrote the book probably can't remember what's there!

How Cancer Works is not a book of lists. In fact, this book is written in the form of "lectures," because I want to talk to you directly, just as if we were together in a classroom. In the first lecture, I give a broad-brush overview of cancer in language anyone can understand. Nevertheless, this is not "baby" cancer: This is the conceptual basis of cancer, told with a minimum of detail and jargon. After the overview lecture, I discuss nine "model" cancers, chosen not only because they are among the most deadly to humans, but also because they offer some of the clearest examples of how cancer works.

For each model cancer, I ask three important questions: What risk factors make it more likely that an individual will get this cancer? What does this cancer do to a person who has it? How can this cancer be treated? As I address these questions, I introduce the biological concepts necessary to understand the answers. So instead of discussing the biology of cancer in one place and the disease itself in another, I unveil the biological concepts exactly at the moment when they make the most sense—when these concepts can be shown to have a direct impact on the disease. For example, you won't find the word "oncogene" mentioned in the overview lecture. I wait until the lecture on leukemia to introduce that term—because chronic myeloid leukemia offers an excellent example of the evil that an oncogene can do. Once you see an oncogene "in action" in leukemia, you won't forget what it is!

This book is intended for anyone—even a non-scientist—who wants to learn about cancer. In college, this book may be used as the text for an introductory course on cancer—but students should keep one point in mind: I didn't write this book for your professor. This book's for you!

An Overview

In the lectures that follow, we will discuss nine "model" cancers. I have chosen these cancers not only because they are among the most deadly, but also because they provide outstanding examples of how cancer works. We will not try to learn everything there is to know about these nine cancers. That would be boring. Rather, we'll spotlight those features of each cancer which best illustrate the biological principles that underlie cancer in general. By the time you finish these lectures, you should have an excellent understanding of the basic features of this disease. However, because we will be learning about cancer in "bits and pieces," it is especially important that we start with an overview. That way, you'll have a feel for where these bits and pieces fit into the overall scheme of things.

CANCER IS A CONTROL SYSTEM PROBLEM

This year, cancer will kill about 600,000 Americans. It is a terrible disease.

Estimated Cancer Deaths in the United States in 2004

Cancer Type	Estimated Deaths
Lung	170,000
Colon/Rectum	55,000
Breast	48,000
Prostate	36,000
Lymphoma	28,000
Leukemia	22,000
Liver	12,000
Skin	10,000
Cervix	5,000
Total (all cancers)	600,000

Cancer arises when, within a single cell, multiple control systems malfunction. These control systems are of two basic types: systems that promote cell growth, and safeguard systems that protect against "irresponsible" growth. Growth-promoting systems determine when, where, and how often a cell will increase in size and divide to produce two daughter cells—a process biologists call proliferation.

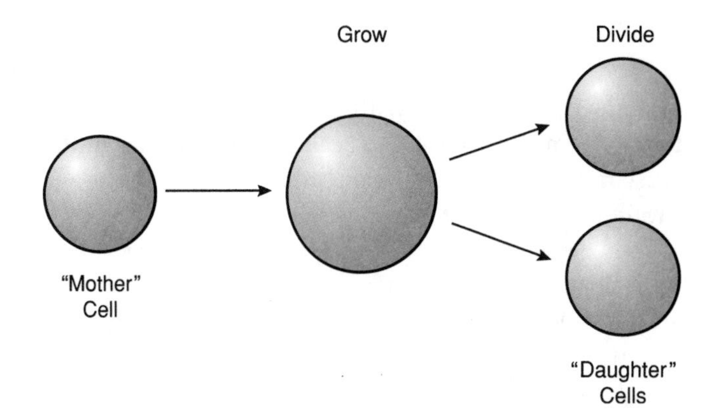

Cell proliferation, when it is controlled properly, is a good thing. After all, an adult human is made up of trillions of cells, so a lot of proliferation must take place between the time we are a single fertilized egg and the time we are full-grown. Once a human reaches adulthood, most cell proliferation ceases. For example, when the cells in your heart have proliferated to make that organ exactly the right size, the heart cells stop proliferating. On the other hand, skin cells and cells that line our body cavities (e.g., our intestines) must proliferate almost continuously to replenish cells that are lost as these surfaces are eroded by normal wear and tear. All this cell proliferation, from cradle to grave, must be carefully controlled to ensure that the right amount of proliferation takes place at the right places in the body and at the right times.

In a sense, the systems in our cells that control proliferation are like the heating system in your home. That system begins with a thermostat which senses the temperature in your house. If heat is required, a switch closes that allows a current to flow down a set of wires to the furnace room. When this electrical signal reaches your furnace, it causes a valve to open, allowing natural gas to enter the combustion chamber of the furnace, where it is burned to produce heat.

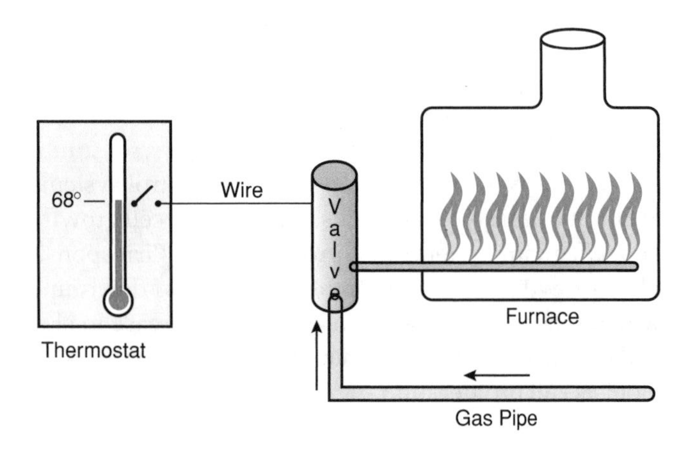

Actually, you may have several of these control systems in your home, with thermostats located in different rooms. Likewise, each cell in our body has multiple control systems that can promote cell growth and proliferation when it is needed.

Most of the time, your heating control systems work just fine, and the furnace comes on only when it should. Occasionally, however, the system may malfunction, so that your furnace ends up being turned on all the time, regardless of the temperature. The same is true of the control systems in our cells. Most of the time they work as intended, but occasionally, one of the growth-promoting systems in one of our cells will malfunction, and the cell will begin to proliferate inappropriately. When this happens, that out-of-control cell has taken the first step toward becoming a cancer cell.

If you look carefully at your face, you may see the result of inappropriate cell proliferation—what doctors call benign growths. The older you get, the more of these you will notice, because as you age, it becomes more and more likely that control systems in one of the many cells on your face will be corrupted. Usually these "blemishes" remain small, because the size of the growth is limited by the distance between the cells that make up the growth and the nearest blood vessel. All of our cells must get their nourishment from blood, and as a result, no living cell in your body is more than about the thickness of a fingernail from a blood vessel.

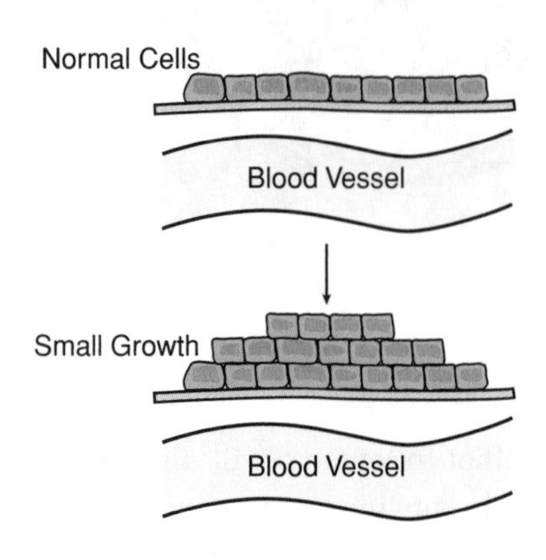

I think most of us have an "Old Uncle Harry" who has a big growth (e.g., a large mole) on his face—one that is much thicker than a fingernail. For a growth to reach this size, new blood vessels are required, because without new blood vessels to provide nourishment, cells that proliferate beyond the range of the normal blood supply will starve to death. So for a wannabe cancer cell to form a mass of any real size, it is not enough just to proliferate inappropriately. Additional systems within the cell must be corrupted so that the wannabe cancer cell produces substances (e.g., proteins) that promote the growth of new blood vessels. These same substances also can be produced by perfectly normal cells. For example, during embryonic development, additional blood vessels are required to supply nutrients to newly formed tissues and organs. Likewise, in response to injury, new blood vessels are needed to replace those which have been damaged. In both cases, substances produced by normal cells orchestrate the growth of the required new blood vessels.

Usually, the systems that control the production of substances which encourage new blood vessels to grow operate normally, wannabe cancer cells are unable to recruit the blood supply they need to continue to proliferate, and they get "stuck" at this early stage in their quest to form a tumor. Every once in a while, however, such a system may go haywire, and a wannabe cancer cell may begin to produce substances which cause new blood vessels to grow. When this happens, the cell can acquire the nourishment it needs to continue to proliferate. That's what happened on Uncle Harry's face.

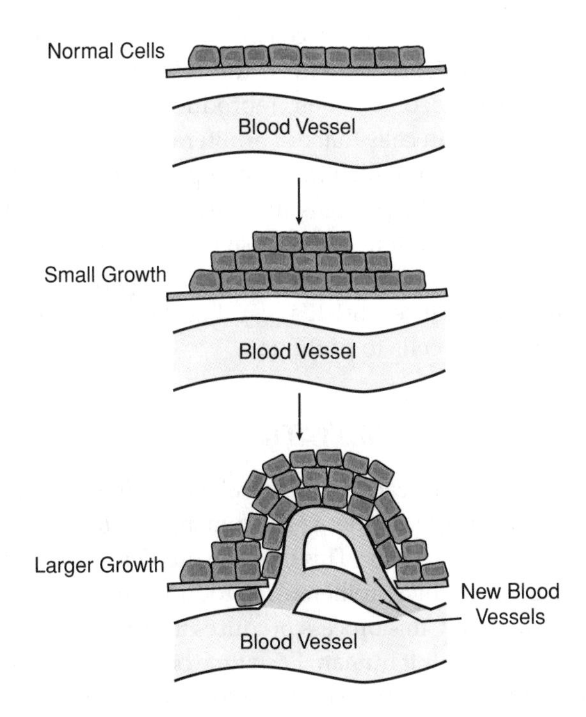

Fortunately, Uncle Harry's growth is probably still "benign," because it grows slowly, and because the wannabe cancer cells that make up the growth have not learned the deadliest trick of all—how to spread to other parts of the body (i.e., how to metastasize). For cancer cells to metastasize, they must first enter the blood stream or the lymphatic system. Scientists still don't completely understand how this happens. They know that normal cells can produce enzymes which break down the tissues that surround them. For example, when tissues are damaged (e.g., because of a wound), normal cells produce enzymes that destroy the damaged tissues, making way for regrowth. As you would expect, the cellular systems that produce these destructive enzymes are under very tight control. However, it is thought that if these systems malfunction, one cell within a previously benign growth may begin to produce enzymes which can destroy the membranes and structures that separate the growth from the blood and the lymph. When this happens, cells can break off from the growth, and can travel with the blood and the lymph to other parts of the body, where they can form secondary tumors. At this point the situation is serious, because although a skilled surgeon can often remove a primary tumor, cancers that have metastasized frequently are fatal.

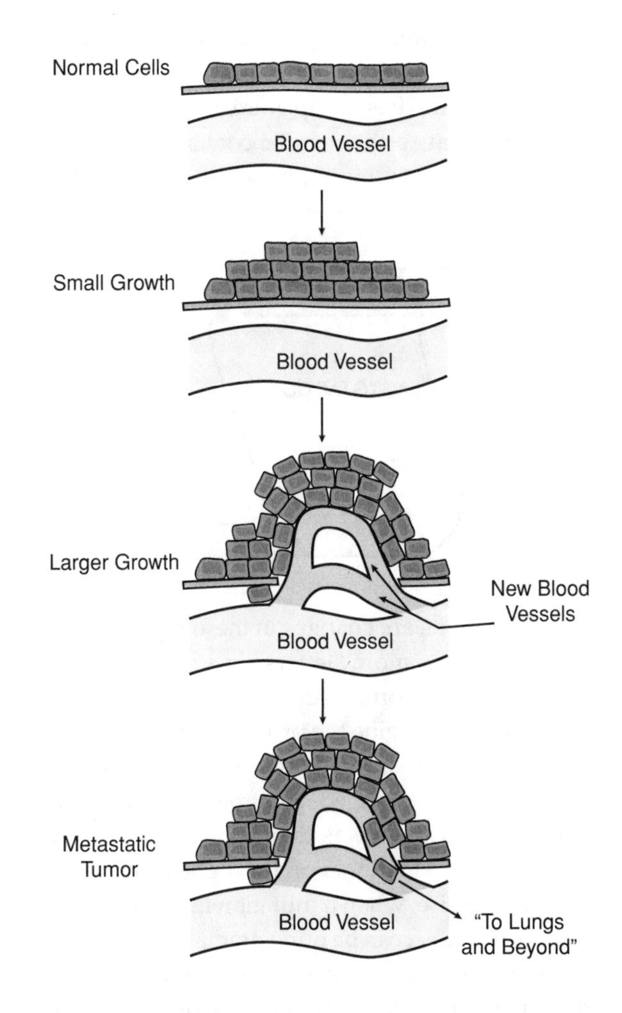

An important point to take from all of this is that a cell really doesn't need to discover anything "new" to become a cancer cell. The control systems that must be corrupted to promote the growth of a cancer cell are normal systems that perform perfectly legitimate, in many cases essential, functions during the life of a cell.

MUTATIONS AND CANCER

So cancer results when systems that control cell growth malfunction. But what causes these malfunctions? Cellular growth-control systems can be corrupted either when cellular genes are mutated or when proteins produced during a viral infection interfere with control system function. Here's how this works.

Located within every living human cell is a collection of "recipe books" called chromosomes which are made of DNA, and which contain the information required to make all the proteins a cell needs to function. These recipes specify not only how each protein is to be constructed using the protein-making machinery of the cell, but also how many copies of each protein should be made. The individual recipes contained within these chromosomes are called genes, and instead of being written in English, these recipes are written using a four-letter code. Because the genetic cookbooks don't have pages, they are really more like scrolls, with each scroll representing one chromosome.

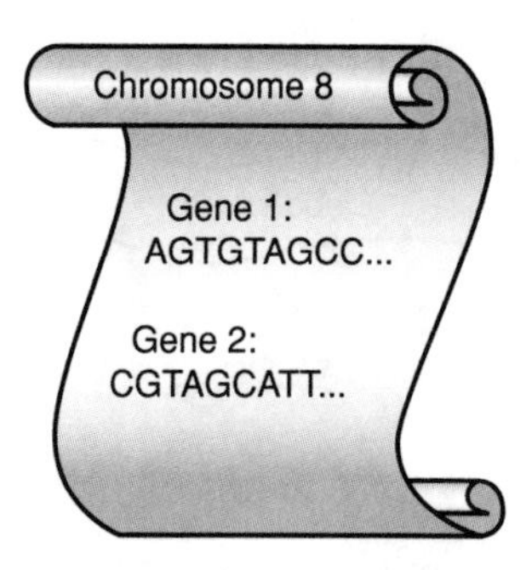

Mutations are changes in these recipes, and they occur when one or more "letters" in a recipe is changed, or when whole "words" or "sentences" are added or deleted. You can imagine the problems that would arise if someone "mutated" a recipe you were using so that the word "beef" was changed to "frog"—or the quantity specified as a teaspoon was changed to a cup. I wouldn't want to be invited to dinner on the night such a mutation occurred! Likewise, when a mutation takes place in a cellular gene, the effect can be quite dramatic. A mutation either can change the character of the protein that is produced (e.g., render it nonfunctional or change its function) or can alter the number of proteins that are made.

Cellular growth-control systems are constructed from proteins. For example, certain proteins act as sensors which help determine whether or not cells should proliferate. Some proteins act as the "wire" that carries the "let's proliferate" signal within the cell, and still other proteins function as switches. Consequently, if a gene for one of the proteins that makes up a growth-promoting system is mutated, that protein may no longer function properly, normal growth control may be lost, and the cell may be triggered to proliferate at a time or place that is not appropriate. Altogether, there are about 35,000 genes in a human cell, but fortunately, only a few hundred of these genes, when mutated, can contribute to transforming a normal cell into a cancer cell. So although mutations in our genes are occurring continuously, we can be thankful that the targets for cancer-causing mutations are relatively few.

In addition to mutant cellular proteins, certain viral proteins can activate the cellular control systems which promote proliferation. Viruses depend heavily on the machinery within the cells they infect to produce new virus particles. However, the cellular "copy machines" they need for their reproduction usually are turned on only in cells that are proliferating. So to create an environment in which they can reproduce efficiently, some viruses make proteins that interact with the cell's growth-control systems and cause inappropriate proliferation. It's not that these viruses want to play a part in making a cell cancerous. It's just that they can survive only by forcing cells to proliferate.

WHAT CAUSES MUTATIONS?

So mutations can lead to uncontrolled proliferation. But what causes these mutations? All humans start life as a single cell—a fertilized egg. This cell then proliferates to make two cells. These then proliferate to produce four cells—and so on. Eventually this process produces the roughly 10 trillion cells in an adult human. Each time one of these proliferating cells gets ready to divide into two cells, its genetic information (in the form of DNA) must be copied, so that the complete genetic cookbook can be passed down to its daughter cells. And each time this copying process takes place, mistakes are made. In fact, on average, about sixty of the DNA letters in the cookbook are changed each time it is copied. This may seem like a lot, but when you realize that the whole cookbook contains about six billion letters, it is clear that the gene-copying process is amazingly faithful. The reason so few errors are introduced during copying is that newly copied DNA is subjected to extensive "spell checking," which detects and corrects most of the errors. Of course, if the process of copying our chromosomes were completely error-free, humans never could have evolved, since mutations are required for evolution. Indeed, it's likely that Mother Nature adjusted the error rate to be high enough to allow for fairly rapid evolution, yet not so high that we would all look like characters out of *Star Wars*.

So the proliferation that takes place during the growth of a human offers the opportunity for copying errors to occur, but of course, all cell proliferation doesn't stop at adulthood. For example, in premenopausal women, hormones produced on a monthly schedule cause cells in the breast and in the lining of the uterus to proliferate. And blood cells proliferate nonstop to produce about 100 billion new blood cells each day. So even when a human is full-grown, there are still many cells in the body that continue to proliferate—and accrue mutations.

In addition to errors introduced during copying of the genetic cookbook, the DNA in our cells is continually being damaged, and this damage can result in mutations. For example, one of the four DNA letters that make up our genes is unstable in the chemical environment of the cell. As a result, this letter is continually being converted (at a low frequency) to another letter, causing a mutation.

Finally, carcinogens that we are exposed to in the environment, toxic chemicals produced in all of our cells as normal byproducts of cellular metabolism, sunlight, manmade ionizing radiation (e.g., x-rays), and the continuous low-dose ionizing radiation from cosmic rays to which every earthling is subjected—all these things can damage DNA and create mutations. These mutations, which alter recipes in the genetic cookbook, can result in defective proteins being produced, and some of these proteins may be involved in controlling cell growth. Consequently, any agent that causes mutations will increase the likelihood that a normal cell will become a cancer cell. That's why activities that induce mutations (e.g., cigarette smoking, working with radioactive materials, or getting sunburned) are "risk factors" for cancer. In fact, although mutations can be inherited from our mothers and fathers, the vast majority of human cancers involve mutations acquired after we were fertilized eggs.

Given the number of mutations that are introduced during the growth of a human, and all the possibilities for suffering mutations once we are adults, the wonder is <u>not</u> that some humans get cancer. The amazing thing is that all humans don't get cancer all the time! In fact, on a per cell basis, cancer is incredibly rare. Even though our bodies are made up of trillions of cells, only one in three humans will get cancer during his lifetime. And it takes only one bad cell to make a cancer.

CONTROL SYSTEMS THAT PROTECT AGAINST CANCER

One reason all of us don't get cancer is that, in addition to growth-promoting systems, our cells also have "safeguard" systems that help protect against inappropriate cell proliferation. These safeguard systems are of two general types: systems that help prevent mutations, and systems that deal with mutations once they occur. For example, our cells have "detox" systems that can convert toxic byproducts of cellular metabolism or carcinogenic substances from the environment into harmless chemicals, preventing DNA damage. In addition, cells have a number of different "repair" systems that can detect damaged DNA and fix it, safeguarding against mutations. These DNA repair systems are especially important, because, on average, the DNA of each of our cells takes about 25,000 mutational "hits" every day! Fortunately, most of this damage is taken care of by the repair systems that safeguard the integrity of our DNA, and only about twenty of these mutations go unrepaired per cell per year.

Although your heating system is designed to operate flawlessly, sometimes things do go wrong. To deal with this possibility, your furnace may be equipped with safeguard systems that help protect your home in case of a heating system malfunction. For example, there may be an "overtemp protector" that will shut down your furnace completely if it runs nonstop and gets too hot. Likewise, because our cells are operating in a mutation-prone environment, Mother Nature wisely has equipped our cells with safeguard systems that can protect against the <u>effects</u> of mutations when they do occur. For example, if the DNA of a cell has been damaged so badly that successful repair would be impossible, there are systems which trigger the cell to die.

The bottom line here is that for a cell to become cancerous, it's not enough just to activate growth-promoting systems. Safeguard systems within the cell must also be inactivated. The current view is that to become a cancer cell, multiple growth-promoting systems must be activated inappropriately, and multiple safeguard systems must be inactivated. Indeed, biologists estimate that about five different control systems usually must be corrupted before a cell becomes cancerous. Since inappropriate activation or inactivation of these systems is caused by mutations, this implies that multiple mutations are required before a normal cell can become a cancer cell. This is the reason why cancer is mainly a disease of the elderly: It generally takes many years for a single cell to accumulate all the mutations required to corrupt these control systems.

Your heating system is composed of multiple elements (e.g., a thermostat, wire, gas valve, etc.), and if any one of these malfunctions, the result is the same: loss of control. Likewise, every growth-promoting or safeguard system is composed of multiple protein elements, and there are many different mutational scenarios that can compromise these systems and transform a normal cell into a cancer cell. In addition, some of the control systems that affect cell growth are unique to certain cell types. For example, the control systems of a blood cell and a lung cell can be very different, because these cells live in different environments, and must respond to different growth signals. Consequently, if we were to learn everything there is to know about leukemia (a blood cell cancer), we would still not understand completely how lung cells become cancerous. Since cancer can arise in more than 100 different cell types, this is a big problem.

CANCER IS A FACT OF LIFE

So cancer is a control system problem. And because the growth-promoting systems that are activated inappropriately in cancer cells are the same systems required for the operation of normal cells, we are stuck with the possibility that one day, these systems will malfunction. Indeed, even if humans lived lives free of any "avoidable" influences that might cause mutations in our DNA (e.g., carcinogens in our food or in the air we breath, diagnostic x-rays, etc.), we would still suffer mutations, and, as a result, would still get cancer. In fact, it is estimated that if everyone lived in a "cancer neutral" environment, there would still be about one-third as many cancers as in the "real world" where many people do stupid things that increase their risk of getting cancer. Yes, cancer is a fact of life: It's a natural consequence of the way our cellular control systems are constructed.

However, before we fault Mother Nature for bad design, we must remember that humans evolved to live only long enough to bear and raise their offspring—usually less than four decades. In contrast, cancer is a disease which generally develops after a person's child-bearing years have passed.

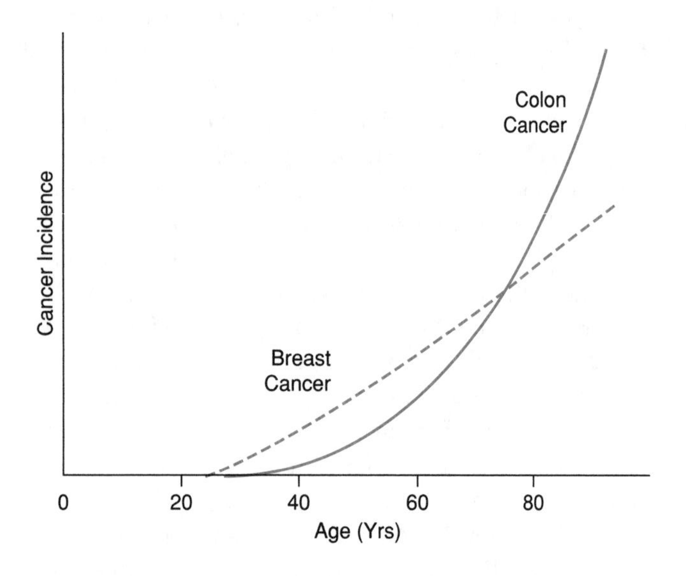

Because most cancers arise in older people, there has been no evolutionary pressure to equip humans with "better" control systems that are less likely to malfunction. Indeed, it could be argued that, because mutations drive evolution, there was actually an advantage in "designing" humans so that a fairly high frequency of mutation was allowed: With a lower mutation rate, evolution would have proceeded more slowly, and all of us might still be bacteria or fungi! My guess is that during evolution, cellular systems were set up so that the mutation rate was high enough to allow rapid evolution, but low enough so that most humans in their reproductive years did not get cancer.

WHAT LIES AHEAD?

In the next six lectures, we will discuss real-life examples of how growth-promoting and safeguard systems are corrupted, often with fatal consequences. Then, in Lecture 8, we will examine the role our immune systems play in dealing with cancer cells. Finally, in the last lecture, we will attempt to look into the future to see what cancer might "look like" a decade or so from now.

Leukemia

Leukemia is a cancer that results when multiple control systems within a maturing blood cell are corrupted. Overall, leukemia accounts for about 3% of cancer deaths in the United States. However, this disease is especially insidious, because it is one of the few cancers that commonly afflicts children. Indeed, leukemias represent roughly 50% of all childhood cancers. In this lecture, we will focus on two types of leukemia: acute myeloid leukemia and chronic myeloid leukemia. I have chosen these as our "models" because they provide especially good examples of the principles that underlie leukemia in general.

WHAT IS LEUKEMIA?

All of our blood cells are made in the bone marrow, where they descend from self-renewing cells called stem cells—the cells from which all blood cells "stem." By self-renewing, I mean that when a stem cell proliferates to produce daughter cells, it does a "one for me, one for you" thing in which some of the daughter cells are "put on reserve" for future use as stem cells, while other daughter cells mature to become red or white blood cells.

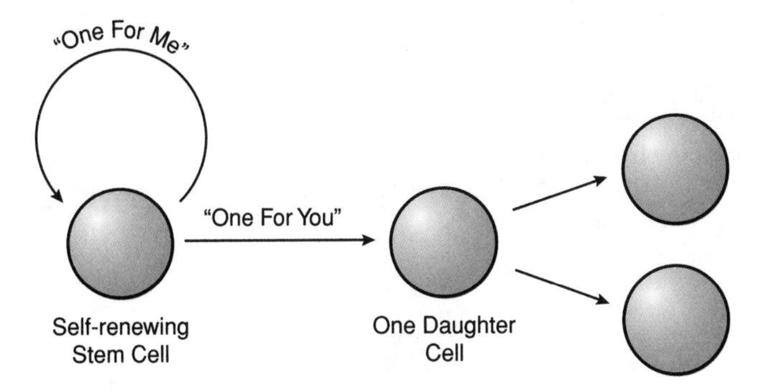

In our bone marrow, there are roughly fifty million blood stem cells. However, experiments with mice suggest that only a small fraction of these stem cells is active at any one time. The descendants of these active stem cells proliferate in the marrow, doubling in number every few days. And as they proliferate, these cells make decisions about the kinds of blood cells they will become when they grow up. Some will mature along the "myeloid" lineage to become macrophages or neutrophils (white blood cells that are important defenders against disease) or red blood cells (which carry oxygen). Others will mature along the "lymphoid" lineage and will end up being T cells or B cells—immune system cells frequently found in lymph nodes. Here is a figure depicting some of the different kinds of blood cells the progeny of a stem cell can become.

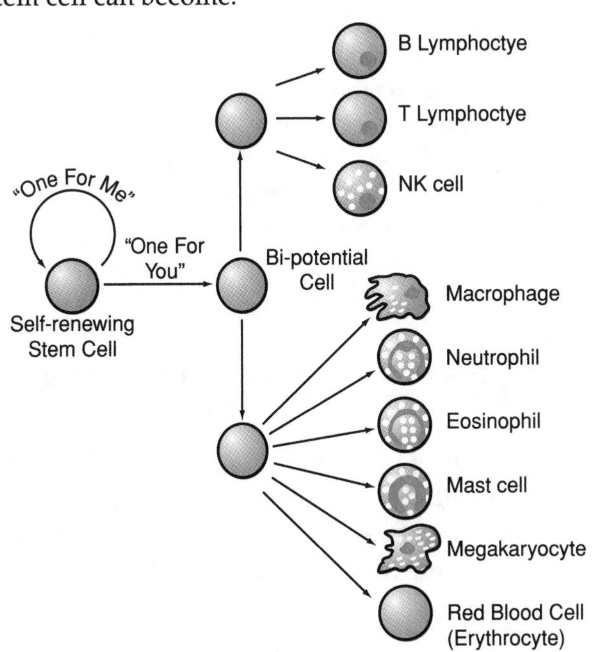

Most of the time this orderly process of proliferation and maturation goes along just fine, and roughly 100 billion blood cells are cranked out each day by the bone marrow. However, on rare occasions, mistakes are made. When this happens, one of the descendants of a blood stem cell may stop maturing, and just continue to proliferate, forming a "clone" of immature cells in the bone marrow. When this happens, the result is leukemia. Because these leukemic cells refuse to grow up, they fill the bone marrow and prevent other blood cells from proliferating and maturing. As a result, the patient usually suffers from anemia due to a scarcity of red blood cells; from infections caused by a deficit of immune system cells; or from hemorrhaging due to lack of platelets (produced by white blood cells called megakaryocytes), which plug holes in leaky or damaged blood vessels.

As blood cells develop, they express certain proteins on their surface that are characteristic both of their type and of their stage in development. By studying these surface proteins, biologists can determine what leukemic cells were on their way to becoming when their development was arrested. Consequently, leukemias can be classified according to the type of blood cell the leukemic cells originally were destined to become. For example, leukemic cells that were progressing toward becoming myeloid blood cells are the culprits in "myeloid" leukemia. Likewise, leukemic cells whose maturation program was arrested as they progressed along the lymphoid lineage give rise to "lymphocytic" leukemia. Leukemias are also classified according to how fast they would kill an untreated patient. In "acute" leukemias, the disease progresses so rapidly that an untreated human usually dies within a few months of diagnosis. In contrast, patients with "chronic" leukemia sometimes live for many years. So leukemia is really a collection of many different blood cell cancers, all of which are caused by blood cells which proliferate in the bone marrow, but don't mature.

In addition to filling up the bone marrow, leukemic cells also can "spill over" into the blood, and leave the marrow. Indeed, leukemia usually is diagnosed by examining circulating blood cells under a microscope to look for cells that have not matured fully. Importantly, it is very difficult to diagnose leukemia by microscopic examination until the leukemic clone has grown to about one billion cells. On the other hand, about two trillion leukemic cells are usually enough to kill a patient. So only about 2,000 times as many cells are needed to kill as are required for a definitive diagnosis. Since in cases of acute leukemia, the leukemic cells usually double their numbers in about three days, and since only about twelve such doublings are required to expand their numbers 2,000-fold, the time between diagnosis and death can be as short as five weeks. This explains why acute leukemias can kill an untreated patient so quickly.

HOW TO BECOME A LEUKEMIC CELL

If you are a cell that "wants to be" a cancer cell, leukemia is your simplest choice. After all, leukemic cells don't have to metastasize to work their evil. They can sit right there in the bone marrow and proliferate until the normal blood cells are "squeezed out." Further, because leukemic cells do not form solid tumors, a wannabe leukemic cell does not have to produce substances which promote the growth of new blood vessels, as a solid tumor (e.g., a lung cancer) must do. So the number of control systems that must be corrupted to turn a normal cell into a leukemic cell are relatively few.

Of course, one control system which must be disturbed to produce a leukemic cell is the system that controls maturation. After all, a defect in blood cell maturation is one of the hallmarks of full-blown leukemia. Although scientists agree on this point, they are not clear on what other controls need to be disabled. Most believe that a system (or systems) which oversees cell proliferation must also be compromised, so that the leukemic cells will have a growth advantage over normal cells in the marrow. Further, it is widely held that if the system which oversees cell maturation is disrupted so that a cell doesn't mature normally, safeguard systems within that

Table 2.1 Different Types of Leukemia and Their Annual Incidence in Adults and Children in the United States

Type of Leukemia	Approximate New Cases: Children (average age)	Approximate New Cases: Adults (average age)
Acute lymphocytic leukemia	2,500 (3)	1,500 (55)
Acute myeloid leukemia	600 (12)	10,000 (70)
Chronic lymphocytic leukemia	0	10,000 (70)
Chronic myeloid leukemia	0	5,000 (65)

cell will sense that something is wrong, and trigger the cell to commit suicide. Consequently, many biologists feel that at least one safeguard system must also be disabled for a cell to survive and become a leukemic cell.

IONIZING RADIATION AND LEUKEMIA

For some time, it has been known that exposure to the "ionizing" radiation which results from the decay of radioactive elements is a risk factor for leukemia. For example, Marie Curie, who received the Nobel Prize for her work with radium at the end of the nineteenth century, died of leukemia. Madame Curie was so intimate with radioactive materials that some of the letters she wrote are still radioactive! Also, as you probably know, many people who survived the atomic bomb attacks on Hiroshima and Nagasaki subsequently died of leukemia.

Experiments have demonstrated that when ionizing radiation interacts with the DNA that makes up our chromosomes, so much energy is imparted that both strands of the DNA molecule can be broken, leaving the attacked chromosome in two pieces.

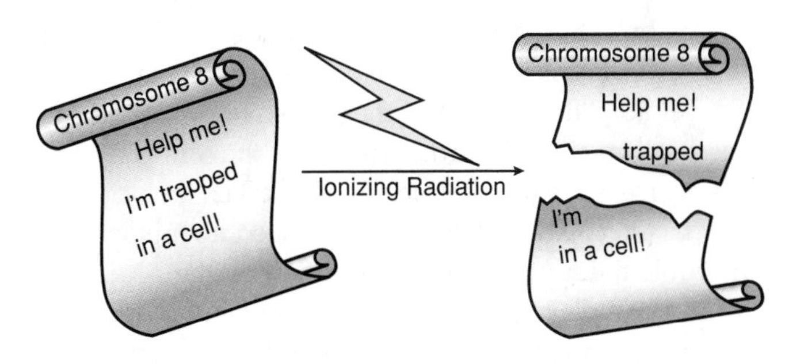

Human cells have evolved two ways of dealing with these double-strand breaks. With the exception of the sex chromosomes, each of our cells has two very similar (but not identical) copies of each chromosome. Consequently, when one chromosome is broken, the unbroken copy of that chromosome can be used as a guide to repair the break. This type of repair (recombination repair) is generally error-free, and is certainly the preferred method for repairing double-strand breaks in chromosomal DNA.

In some cases, another strategy is used to repair double-strand breaks. It turns out that the ends of broken chromosomes are quite "sticky," and sometimes, before recombination repair can be carried out, the broken ends of a chromosome are simply glued back together. This type of repair (end joining) usually results in the loss or gain of a few of the building blocks from which the DNA molecule is constructed, causing a mutation.

Occasionally more than one chromosome may be broken within the same cell. When this happens, end joining can result in part of one broken chromosome being mistakenly joined to part of another broken chromosome. This type of mutation is called a translocation, and it can lead to the disruption of genes located at the spots where the two chromosomes were broken. Translocations can also cause parts of two different genes to be "fused" together, creating new genes.

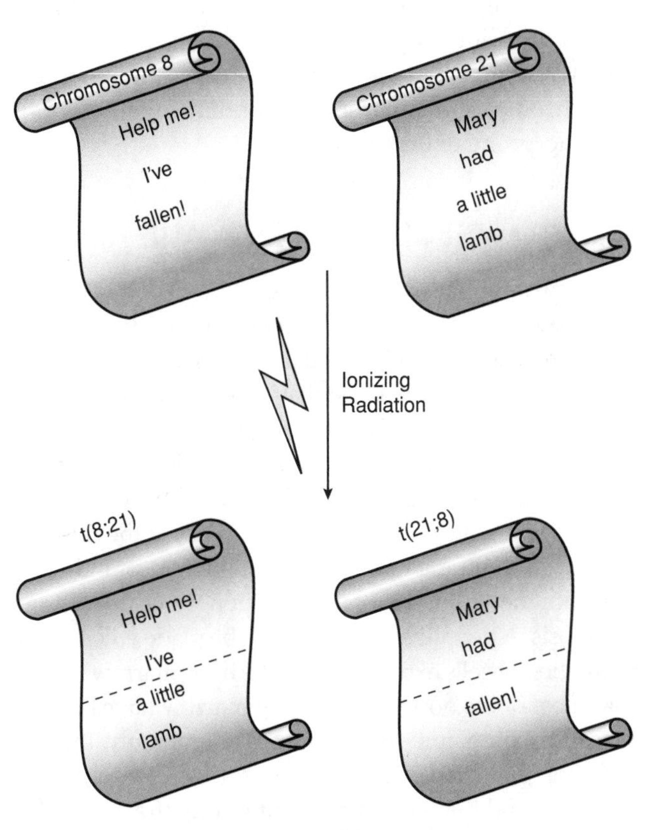

Translocated Chromosomes

Because translocations are relatively easy to spot (most translocated chromosomes look funny under a microscope), because ionizing radiation can cause translocations, and because persons who are exposed to high levels of ionizing radiation frequently get leukemia, biologists examined the DNA within the leukemic cells of patients who had been exposed to ionizing radiation to see if there were any "common" translocations. The thinking here was that if the same chromosomal translocations were found in the leukemic cells of many different patients, the places where the chromosomes were broken might pinpoint the locations of genes that were involved in leukemia.

The search for common translocations was successful, and a number of translocations were identified that occurred frequently in cases of radiation-associated leukemia. Flushed with success, biologists also examined the chromosomes of leukemia patients who had not been exposed to high level ionizing radiation—and they found something very surprising. These leukemia patients also had translocations, and frequently these translocations were the same as those found in the leukemic cells of patients who had been exposed to ionizing radiation! Biologists still aren't sure what causes all these translocations, but it appears that chromosomes in blood stem cells are rather "fragile" and prone to breakage.

ACUTE MYELOID LEUKEMIA AND AML1

One translocation that results from the fusion of part of chromosome eight to part of chromosome twenty-one, usually written as t(8;21), is detected in about 15% of all cases of acute myeloid leukemia—the most common leukemia in adults. As the name implies, the leukemic cells in individuals with acute myeloid leukemia were destined to become cells of the myeloid lineage—for example, a neutrophil. Acute myeloid leukemia is one of the most aggressive human malignancies. Untreated, most individuals will die from this cancer within three months of diagnosis. A definitive diagnosis can be made when an examination of cells extracted from the bone marrow reveals that most of the normal cells in the marrow have been replaced by immature leukemic cells. Commonly, the first signs of this disease include weakness, bleeding, easy bruising, and susceptibility to infections (frequently accompanied by fever). These symptoms are the result of the decreased production of normal red blood cells (causing weakness due to anemia), platelets (causing bleeding and bruising), and immune system cells (inviting infections).

When the DNA sequence surrounding the t(8;21) breakpoint in acute myeloid leukemia was analyzed (not an easy task!), it was found that this translocation fuses parts of two genes together. One of these genes was named AML1 (for obvious reasons) and the other was called ETO (which stands for eight twenty one). Once these genes were identified, researchers sprang into action to determine both their functions in normal cells and their defects in leukemic cells. It is still not known what the normal function of the ETO gene is, but studies of the normal AML1 gene have yielded very interesting results.

To get an idea of what the AML1 gene does in a normal cell, biologists engineered mice that lack the AML1 gene. I won't go into how they did this, but you can understand, I'm sure, that it was a real piece of work. Unfortunately, the mice without the AML1 gene died before they could be born. But they died in a very informative way. Their blood cells did not mature! And, of course, without functional blood cells (e.g., red blood cells to carry oxygen), mice cannot survive gestation. So these experiments demonstrated that the normal AML1 gene specifies a protein that is absolutely critical for the maturation of blood cells in mice. I say "in mice" because one must be careful when applying lessons learned in mice to humans, since some of the control systems in mouse cells operate differently from those in human cells. After all, humans are not just big mice.

Further experiments revealed at least one way AML1 is involved in blood cell maturation. In a human cell, the protein specified by the AML1 gene has a partner protein (CBFβ) to which it binds. When these two proteins get together, they can attach to the cell's DNA at specific "control regions," and turn on the expression of various cellular genes. Proteins that are able to regulate the levels at which cellular genes are expressed are called transcription factors. One of the genes that is under AML1's influence (C/EBPα) is known to be crucial for the maturation of myeloid cells. Taken together, these findings indicate that the normal AML1 protein is part of the control system that oversees the maturation of blood cells.

Of course, biologists also need to understand what role the mutant protein specified by the AML-ETO fusion gene plays in acute myeloid leukemia. Although there is still much to be learned about the function of this protein, I'll tell you what has been discovered so far. It turns out that when the AML1 gene is mutated by being fused to part of the ETO gene, the mutant protein specified by this fusion gene can still bind to the CBFβ partner protein. But when it does, this partnership is unproductive: The duo no longer functions as a transcription factor that regulates the expression of genes involved in blood cell maturation. So the current view is that in leukemias that have the 8;21 translocation, the breaking and joining of chromosomes produces a mutant gene which encodes an AML-ETO fusion protein. Although these cells still have one good copy of chromosome 21 which can direct the production of normal AML1 proteins, the mutant AML-ETO fusion protein binds to and "soaks up" most of the available CBFβ partner protein. Consequently, there are not enough "good" AML1/CBFβ partnerships formed to cause the leukemic cells to mature. The bottom line is that as a result of the AML-ETO translocation, the system that controls the maturation of blood cells is disrupted.

Because blood cells with the 8;21 translocation don't mature, they possess one of the evil qualities of leukemic cells—arrested development. I say one of the qualities because this translocation alone is not sufficient to cause leukemia. For example, when the gene for the AML-ETO fusion protein is artificially expressed in normal mouse cells, these modified cells do not act like full-blown leukemic cells. This suggests that although the AML-ETO translocation causes a malfunction in the system that controls blood cell maturation, to become a full-blown leukemic cell, other control systems must be disrupted. This view is consistent with the observation that the peak time for diagnosis of radiation-induced acute myeloid leukemia is about ten years post exposure. Presumably, this much time is required to accumulate the additional mutations needed to cause other control systems to malfunction. Indeed, in humans, no single mutation has been discovered which alone can cause cancer. Multiple control systems must be compromised to turn a normal cell into a full-blown cancer cell. This is true even in leukemia—a cancer in which relatively few control systems must malfunction for a blood cell to become cancerous.

The fact that the AML-ETO mutation alone is not enough to turn a normal blood cell into a leukemic cell—that additional mutations are required—raises an interesting question. Since it generally takes years for a cell with the AML-ETO translocation to pick up the "extra hits" required to produce a full-blown leukemic cell, why don't cells which undergo the 8;21 translocation die or leave the marrow before this happens? After all, the life span of most blood cells is only days or weeks. The answer to this question isn't known for sure, but many biologists feel that the only way a blood cell could possibly hang around long enough to accumulate the required multiple mutations is if the cell accruing these mutations were one of the stem cells—not one of its daughters or granddaughters. The reason is that as the stem cell does its one-for-me-one-for-you thing, any mutations that have occurred during the life of the stem cell not only will be passed down to its progeny, but also will be retained by the stem cell as it self-renews. According to this view, bone marrow stem cells represent a reservoir for the accumulation of mutations, some of which may lead to leukemia.

TREATMENT OF ACUTE MYELOID LEUKEMIA

Patients with acute myeloid leukemia whose leukemic cells show the 8;21 translocation usually receive "induction" therapy. This treatment requires about a month of hospitalization and costs about $100,000. During treatment, the patient is subjected to almost lethal chemotherapy, using drugs (e.g., daunorubicin and cytarabine) which are designed to destroy most, but not all of the cells in the patient's bone marrow. Ironically, these drugs work by damaging DNA, and take advantage of the fact that the leukemic cells are proliferating, whereas most other cells in the body are not. The effect of DNA-damaging agents is more severe in proliferating cells than in cells that are not proliferating (e.g., other stem cells that are not currently "in use"), because proliferating cells have less time to repair their damaged DNA, and often die as they try to produce daughter cells. Most of the drugs currently used to treat cancer target proliferating cells.

Of course, there are some normal cells in the body that are proliferating. For example, cells that line the digestive tract are constantly being renewed by proliferation, so these cells are very sensitive to chemotherapy. That's one reason why patients receiving chemotherapy frequently suffer from nausea and diarrhea. Also, since most of a patient's blood cells are destroyed by the treatment, persons undergoing induction therapy must be isolated in a protective environment (because their immune systems are impaired) and must receive regular platelet and red blood cell transfusions.

About half of the young and middle-aged adults who undergo induction therapy for acute myeloid leukemia survive for more than four years. For older people, the percentage of cures using induction therapy is smaller. The reason for the relative lack of success using chemotherapy on older patients is not well understood. However, experiments in mice have shown that in young mice, less that 10% of their blood stem cells are "in action" at any given moment. In contrast, in older mice, almost all of their stem cells are proliferating. If this result holds true in humans, it would suggest that the nonleukemic stem cells of older humans may be more easily damaged by chemotherapy because they are more likely to be proliferating at the time of treatment.

We frequently hear that people who have undergone chemotherapy for leukemia are in "complete remission." This means that the number of leukemic cells has been reduced to a level such that none can be found. How good this news is depends, of course, on how few leukemic cells could be detected. Using visual techniques, in which a microscope is employed to look for immature blood cells, the limit of detection is about one billion leukemic cells total in the body of the patient.

Consequently, a patient who has fewer than one billion leukemic cells in his body is said to be in complete "cytologic" remission. Using the most sensitive techniques available—techniques such as the polymerase chain reaction that can identify leukemia-associated genetic mutations—leukemic cells can be detected when the patient has as few as 100,000 total leukemic cells. A patient is said to be in complete "cytogenic" remission when no leukemic cells can be found using these very sensitive methods. The bottom line here is that being in remission doesn't necessarily mean that the patient is disease-free. It simply means that no leukemic cells can be detected.

CHRONIC MYELOID LEUKEMIA AND THE PHILADELPHIA CHROMOSOME

Certainly the most famous example of a chromosomal translocation associated with leukemia is the "Philadelphia chromosome" (named after the city in which this mutation was first discovered). The Philadelphia chromosome results from the breakage and joining of chromosomes nine and twenty-two to form a BCR-ABL fusion gene made from part of a gene called BCR on chromosome twenty-two and part of the ABL gene on chromosome nine. This translocation is present in the blood cells of virtually all patients who suffer from chronic myeloid leukemia—a cancer which accounts for about 15% of the cases of adult leukemia. Interestingly, the Philadelphia translocation, discovered in 1959, was the first chromosomal abnormality identified that is consistently found in a human cancer.

Chronic myeloid leukemia (CML) is a very instructive cancer because it actually has two phases. The first or "chronic" phase is characterized by the overproduction of cells of the myeloid lineage (primarily neutrophils), resulting in an imbalance in the number of blood cells of each type that is produced in the marrow. Importantly, these "leukemic" cells are completely functional (i.e., they mature fully). Consequently, because leukemia is usually defined as being a disease in which there is an overproduction of non-functional, immature cells, it could be argued that the chronic phase of chronic myeloid leukemia is really not "full-blown" leukemia.

The overabundance of myeloid cells usually produces white blood cell counts that are extremely high, and about half of the cases of chronic myeloid leukemia are diagnosed during a routine blood test (e.g., as part of a yearly physical exam). In some cases, the abnormally large number of blood cells causes an enlargement of the patient's liver or spleen—two organs in which blood cells accumulate. Because the Philadelphia chromosome is found in essentially all cases of chronic myeloid leukemia, the diagnosis can be confirmed by looking for this translocation.

Clues as to the role the Philadelphia chromosome plays in chronic myeloid leukemia have come from examining what happens when the BCR-ABL mutant gene is introduced into normal cells. Although the mutant BCR-ABL protein clearly has multiple effects on these cells, there is one function of the mutant protein that is likely to be important in chronic myeloid leukemia: The BCR-ABL protein interferes with the workings of a cellular system that senses the presence of growth-promoting proteins outside the cell, and triggers cell proliferation. Such control systems are referred to as growth factor pathways. Here's how they work.

GROWTH FACTOR PATHWAYS

As we discussed earlier, after blood stem cells proliferate, their progeny must "decide" what types of cells to become. As you can imagine, these choices are carefully controlled to make sure we have enough of each kind of blood cell. The latest thinking on how this is accomplished is that as a blood cell matures, the initial choice of the kind of cell it will eventually become is made at random. It's as if a cell just rolls the dice and picks an occupation based on the result. Or at least that's how it appears. Once this decision has been made, the cell begins to display molecules on its surface called growth factor receptors. Importantly, cells that have chosen different fates (i.e., to become different types of blood cells) express different combinations of these receptors. Growth factor receptors function as "antennae" which detect the presence of proteins called (you guessed it!) growth factors. And when growth factors bind to their favorite receptors, cells that have these particular receptors are triggered to proliferate (i.e., to "grow").

For example, cells that have chosen to become red blood cells express erythropoietin receptors on their surface. Erythropoietin is a growth factor that is produced in the kidneys. When kidney cells sense that more red blood cells are needed (because oxygen levels in the kidney are too low), they crank out more erythropoietin. This "extra" erythropoietin then travels with the blood to the bone marrow, binds to erythropoietin receptors on maturing red blood cells, and tells them to proliferate. In this way, the number of red blood cells produced can be adjusted so that there are just the right number of these cells available to carry oxygen throughout the body. Interestingly, the erythropoietin protein can now be produced in the laboratory, and this man-made growth factor can be used to increase the production of red blood cells in patients who are anemic (frequently as a result of cancer chemotherapy).

Other types of blood cells have different types of receptors which bind to different growth factors. But the idea is the same: The number of blood cells of each type that is produced is determined mainly by how much these particular blood cells proliferate—and how much they proliferate is determined by how much of their particular growth factors are present in the marrow. By adjusting the "cocktail" of growth factors that is available, the required number of each blood cell type can be maintained or adjusted to suit the conditions (e.g., during an infection or after a hemorrhage has occurred).

HOW GROWTH FACTORS WORK

But how does the binding of a growth factor to its receptor on the surface of a cell tell that cell to proliferate? After all, these growth factors act outside the cell, yet the genes that must be turned on or off to cause proliferation are located way inside the cell in the nucleus. You see the problem.

Although there are variations on the theme, growth factor receptors usually work in the following way. Growth factor receptors are long proteins that have a "head" which protrudes outside the cell, a "body" which spans the cell membrane, and a "tail" which extends inside the cell into its cytoplasm.

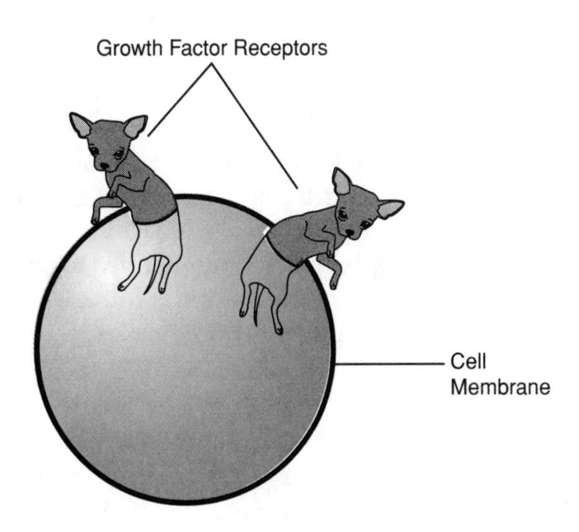

This three-part structure has been recognized for some time, but biologists were at a loss to figure out what a growth factor might do to the head of its receptor that would somehow cause the tail of the receptor to send the "start proliferating" signal to the cell nucleus. There were lots of interesting ideas suggested to try to solve this riddle. Some proposed, for example, that a growth factor might work by grabbing the head of the receptor and pulling it out a bit, and that this might somehow change the shape of the tail part, allowing it to function. This was a nice try, but it turned out not to be correct.

The first piece of information that helped solve the mystery of how growth factors signal was the discovery that growth factor receptors usually cannot send the proliferation signal unless they are paired up. This led to the understanding that growth factors function by grabbing two receptor molecules and bringing them together. Indeed, the current picture is that when a growth factor brings two growth factor receptor molecules close together on the surface of a cell, the "start proliferating" signal is sent.

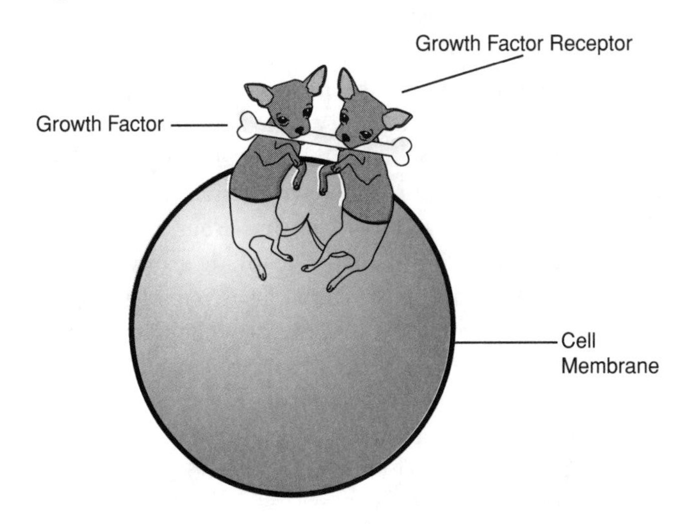

Okay, so growth factors "cluster" their receptors, but how does this send a signal? The answer to this question emerged from experiments which showed that the tail portion of your standard growth factor receptor actually can function as an enzyme called a kinase, which can add a phosphate group to another protein—but not to just any old protein. Indeed, the kinase enzymes within a cell are very picky about what proteins they add a phosphate group to and where on that protein they attach it. In the case of the growth factor receptor kinase, the target protein just happens to be the tail of the other paired growth factor receptor. So when a growth factor gathers together two of its receptor proteins on the cell surface, the tails of these receptors are brought close enough together inside the cell to allow the tail of each growth factor receptor to add a phosphate group to the tail of the other growth factor receptor—the molecular equivalent of mutual back scratching.

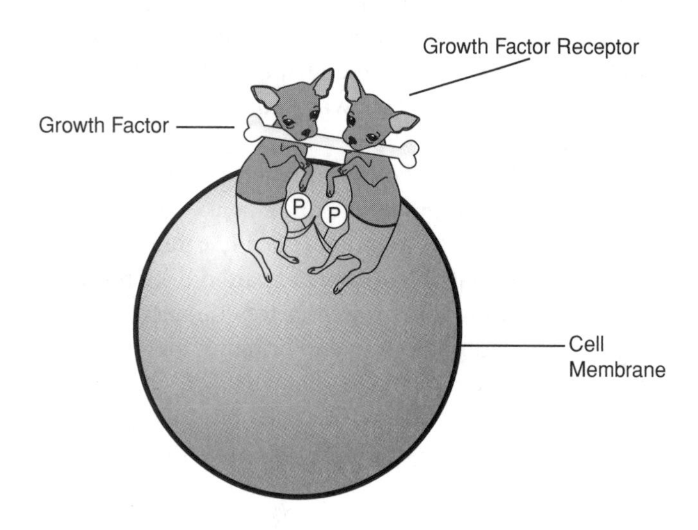

Although a protein can be envisioned as a string of beads (the beads being the amino acids that make up the protein), this string of beads is never straight. In fact, a lot of how a protein functions is determined by the way the protein folds up around itself and around other molecules with which it interacts. Phosphate groups have a negative charge, and electric charge is one of the most important determinants, not only of what shape a protein will take, but also of how that protein will interact with other molecules. So sticking a phosphate group on a protein can change both its shape and its function. The act of attaching a phosphate group to another molecule is called phosphorylation.

Once the tails of the growth factor receptors have been phosphorylated, they are able to initiate a cascade of events that carries the "start proliferating" signal right into the nucleus. This cascade can be thought of as a "phosphorylation relay." The basic rules of this "race" are that each protein in the relay adds a phosphate group to the next protein, which is then able (because of the added phosphate) to function as a kinase to phosphorylate the next protein in the relay. Finally, transcription factors in the cell nucleus get the "handoff" of a phosphate group, which changes their shapes so that they can turn on or off the genes needed for cell proliferation. It's really amazing that the simple addition of phosphate groups makes "signal transduction" (literally, the carrying over of a signal) possible.

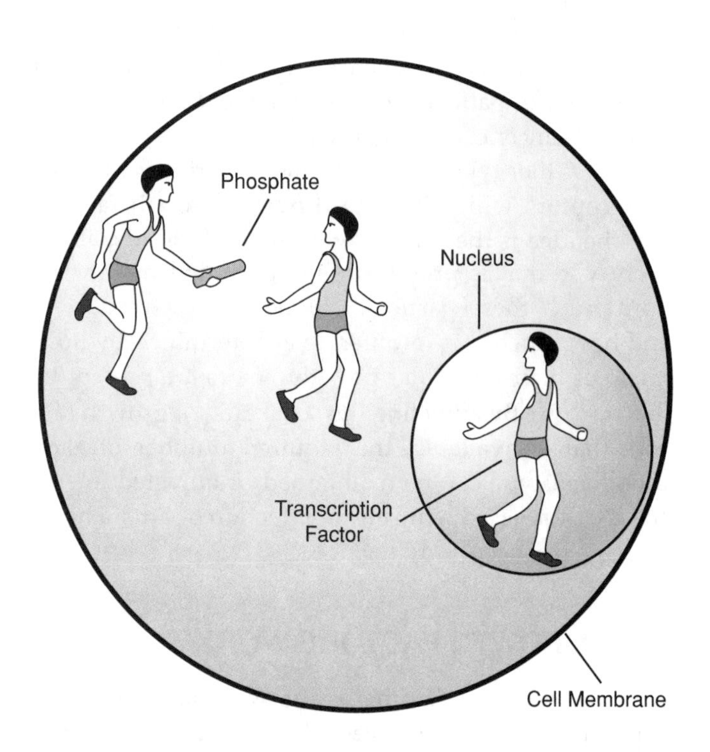

THE BCR-ABL PROTEIN SHORT-CIRCUITS A GROWTH FACTOR PATHWAY

No, I haven't forgotten that we were discussing the mutant BCR-ABL gene and leukemia. And here's where all this growth factor receptor business comes in. It turns out that the ABL protein is a member of a growth factor relay, and normally the function of the ABL kinase enzyme is carefully regulated: The ABL protein only sticks a phosphate group on the next protein in the relay when it is appropriate. However, if the ABL gene is broken and joined to the BCR gene, the shape of the ABL protein is changed, and the regulation of its enzymatic function is lost. The result is an "always on" kinase that functions in a growth factor pathway, independent of whether the growth factor receptors are engaged by growth factors. This situation is analogous to what would happen if the wires from your thermostat were shorted together, sending the "I need heat" signal to the furnace without regard for the temperature in your house.

So the mutated BCR-ABL fusion protein "short-circuits" a growth factor pathway. Consequently, cells with the BCR-ABL mutation do not require growth factors (or not as many growth factors) to trigger proliferation—because the ABL kinase always signals the next runner in the relay to go. As a result of this reduced dependence on growth factors, blood cells with the BCR-ABL translocation proliferate more rapidly than their growth factor-regulated neighbors. After a while, they fill up the bone marrow, causing the symptoms associated with the chronic stage of chronic myeloid leukemia. Indeed, patients with chronic myeloid leukemia typically have ten to twenty times the normal number of white blood cells.

ONCOGENES

The mutant BCR-ABL gene is an excellent example of what biologists call an oncogene. A cellular oncogene is a normal gene which has been mutated so that the protein it produces plays an active role in transforming a normal cell into a cancer cell. For example, the ABL gene is a perfectly normal gene that specifies a protein which is an important component of a growth factor relay. However, when the ABL gene is mutated by being broken and fused to the BCR gene, the resulting mutant ABL protein (called an oncoprotein) functions in an uncontrolled manner to trigger inappropriate cell proliferation. The normal ABL gene is called a proto-oncogene, because it has the capacity, when mutated, to become an oncogene. Fortunately, the vast majority of our roughly 35,000 genes are not involved in promoting cell growth. We can be thankful for that, because proto-oncogenes are "accidents waiting to happen."

The situation with BCR-ABL illustrates another important point: Proto-oncogenes are genes that have normal (frequently indispensable) functions within the cell. Indeed, mice which have been engineered to lack the normal ABL gene die within a couple of weeks of birth and have abnormal blood cells. So although cellular proto-oncogenes are potentially dangerous, a cell really has no choice but to deal with the possibility that a proto-oncogene will go wrong. Said differently, because of the way the control systems in our cells are set up, we are stuck with the possibility that at some time during our lives, one of our proto-oncogenes may be mutated to become a deadly oncogene. This risk is just part of being human.

Each of our cells is equipped with two copies of the ABL proto-oncogene, because each cell has two copies of chromosome nine where this gene resides. When the 9;22 translocation occurs, only one copy of the ABL proto-oncogene is mutated. The other copy is left "proto." But this doesn't matter. One bad copy of the ABL gene is enough to short-circuit the kinase relay and cause a cell to proliferate. The same is true of your heating system. In your house, you probably have several thermostats and their associated control systems, yet only one of these thermostats needs to be switched on for your furnace to fire up. And that's why proto-oncogenes are potentially so dangerous: Only one copy of a proto-oncogene need be mutated for a growth-promoting system to be compromised.

PROGRESSION OF CHRONIC MYELOID LEUKEMIA

Untreated, chronic myeloid leukemia always progresses from a chronic stage to an acute stage. It probably would have been better to call this disease chronic/acute myeloid leukemia, but I'm afraid we are stuck with the current terminology. What is especially interesting about this type of leukemia is that progression from chronic to acute always takes place—and that it usually happens within a short period (typically about five years). During the chronic phase, cells with the Philadelphia chromosome mature normally, but there are just way too many of them. Later, however, when these cells progress from the chronic to the acute stage, they continue proliferating, but they stop maturing. Consequently, leukemic cells in the acute phase of chronic myeloid leukemia have both hallmark properties of full-blown leukemic cells: failure to mature and uncontrolled proliferation.

Oncologists have been puzzled for many years by the rapid transition from chronic leukemia to deadly acute leukemia. The thinking, of course, is that during their five or so years in the chronic stage, leukemic cells accumulate additional mutations that block their maturation. However, the real question is: How can the additional mutations required for disease progression possibly be acquired in 100% of the cases within such a short time span? It just seems too improbable.

New experiments suggest a solution to this conundrum. Because mutations can cause big problems, human cells have evolved multiple safeguard systems for repairing damaged DNA. Recently, it was discovered that one of the most important of these repair systems (the DNA-PK system) is deficient in cells that have suffered the 9;22 translocation. Preliminary results indicate that the BCR-ABL mutant protein somehow causes the destruction of an important component of the DNA-PK repair system. If proven true, this would mean that cells with the Philadelphia chromosome are more "mutation-prone" than normal cells, because an important safeguard system which repairs damaged DNA has been disabled. This could explain why cells of patients with chronic myeloid leukemia accumulate the mutations needed to become full-blown leukemic cells so rapidly.

As we have already discussed, "real" leukemia cells have two major characteristics: unrestricted growth and failure to mature. Because the BCR-ABL mutation leads to uncontrolled cell growth, it would be a safe bet that at least one of the mutations involved in the transition from the chronic to the acute phase of chronic myeloid leukemia is in a system that controls whether or not cells mature. Indeed, the 8;21 translocation, which prevents cells from maturing in acute myeloid leukemia, also has been detected in the leukemic cells of some patients who have transitioned from the chronic to the acute phase of chronic myeloid leukemia. So far, it is not known whether the combination of the 9;22 translocation and the 8;21 translocation is all that is needed to change a normal blood cell into a full-blown myeloid leukemia cell—or whether additional mutations in other control systems are required.

TREATMENTS FOR CHRONIC MYELOID LEUKEMIA

In some cases, the chronic phase of chronic myeloid leukemia can be "managed" by treating patients with interferon alpha—although doctors still don't understand how this drug works to decrease the number of leukemic cells. For patients who do not respond to this drug, three approaches currently are being employed to treat chronic myeloid leukemia: standard chemotherapy, bone marrow transplantation, and gene-specific chemotherapy.

Standard Chemotherapy

Standard chemotherapeutic drugs like hydroxyurea damage DNA and kill off some of the rapidly proliferating blood cells. In a patient whose disease is managed by interferon or standard chemotherapy, the chronic phase can sometimes last for twenty years or more, but most frequently for only about five years. In the latter part of the chronic phase (sometimes called the accelerated phase), patients begin to experience fever, weight loss, and bone pain as they progress to the acute phase or "blast crisis." Blast is a word biologists use to describe immature cells that proliferate rapidly. Consequently, the terms "blast crisis" and "full-blown leukemia" really are equivalent. Patients in the acute phase can be treated with standard chemotherapeutic drugs, but their response to these treatments is generally unsatisfactory, and most succumb within three to six months. So although chemotherapy can help patients with chronic myeloid leukemia live longer, and can improve the quality of their lives, chemotherapy cannot cure chronic myeloid leukemia.

The reason it is impossible to cure this disease with standard chemotherapy is that at doses sufficient to kill all the leukemic cells, the normal stem cells would also be wiped out. The errant cells in the chronic stage of chronic myeloid leukemia are growing only marginally more rapidly than are the other blood cells. Consequently, the leukemic cells and the normal stem cells aren't all that different—and if you kill one, you kill the other. By the time a patient has progressed to blast crisis, when his leukemic cells are growing rapidly, these cells usually have suffered additional mutations which make them resistant to the effects of chemotherapeutic drugs.

So if an oncologist were to try to cure a patient with chronic myeloid leukemia by using high-dose chemotherapy, he might succeed in killing all the leukemic cells, but the patient would be left with no blood cells. The patient could be given transfusions of red blood cells (to carry oxygen) and platelets (to keep him from bleeding to death), but this would get old very quickly. Even worse, the patient would have no functioning immune system cells. This problem is especially acute because neutrophils, which are very important in dealing with everyday infections, have a life span of only a few days, so long-term neutrophil transfusions would be out of the question.

The obvious way out of this dilemma would be to treat the patient with high-dose chemotherapy to destroy all the leukemic cells, and then give him a "stem cell transfusion." After enough time, the transfused stem cells might proliferate, mature, and renew the patient's blood system. Indeed, the reconstruction of a patient's blood system is exactly what a bone marrow transplant is designed to accomplish.

Bone Marrow Transplants

I remember when I first heard about bone marrow transplants. I imagined that a surgeon had to cut a bone out of the donor and graft it into a bone of the recipient. Clearly, only superheros would ever consider being donors for bone marrow transplants! Fortunately, it doesn't work like that. Under a general anesthetic, a surgeon uses a syringe with a large needle to pierce the bones of the donor and to remove a total of about a pint of bone marrow. Marrow has a consistency similar to blood, so it's rather easy to suck this much out of the interior of large bones. Next, the donated marrow is infused into the blood stream of the patient—and then something truly amazing happens: The stem cells from the infused marrow find their way "home" to the patient's bone marrow! In fact, experiments with mice indicate that about 1% of the stem cells that are infused into the blood stream will end up in the bone marrow and begin proliferating there. This process is called engraftment, and the migration of stem cells to the marrow is thought to be controlled by Velcro-like molecules on the surface of the stem cells that stick to other Velcro-like molecules that are present in the marrow.

Although they can be life saving, bone marrow transplants are a rather perilous procedure. First, engraftment is not always successful, so even this first step can be problematic. Second, it usually takes four or five months for the stem cells in a bone marrow transplant to rebuild the immune system of a patient whose blood system has been "deleted." During this time, the patient is susceptible to infection, and he usually must spend some time in a "germ-free" environment while his immune system recovers.

Matching MHC Proteins

There is another major problem with bone marrow transplants: finding a perfect match. Humans have proteins on the surface of their cells called major histocompatibility complex proteins (MHC for short). These MHC molecules give our cells a distinctive "fingerprint." For example, every human has three genes for MHC molecules called HLA-A, HLA-B, and HLA-C that are located on chromosome six. Because we have two chromosome sixes (one inherited from Mom and one from Dad), each of us has a total of six of these genes. In the human population, there are many different variants of the MHC genes. For example, there are at least 125, slightly different forms of the HLA-A gene and more than 260 slightly different HLA-B genes. Because of this diversity, the probability that your MHC "fingerprint" will match mine is very small. Indeed, it is estimated that if you had access to a bank of bone marrow contributed by 10,000 different individuals who were not related to you, the chance that you would find a match to your HLA-A, HLA-B, and HLA-C genes would only be about 70%. Your MHC fingerprint can be analyzed by drawing some of your blood, and using antibodies that recognize various types of MHC proteins to determine which MHC molecules are present on the surface of your cells. If a more detailed fingerprint is required, the genes that specify your MHC proteins can be sequenced. This procedure will reveal tiny variations in your MHC genes that might have been missed by the antibodies.

It turns out that when bone marrow is transplanted, not matching MHC molecules can lead to serious problems. Bone marrow contains, in addition to newly made blood cells, mature immune system cells called killer T cells. These killer cells are extremely important in protecting us against viral infections, because they can kill virus-infected cells before the viruses have time to multiply inside them. However, these same killer T cells have a property that is very annoying to oncologists—they simply hate cells which have MHC molecules on their surface that are different from their own. In fact, when they see a cell with "foreign" MHC molecules, they kill it. So if a donor's MHC molecules are different from those of a patient who is receiving a bone marrow transplant, the killer T cells from the bone marrow of the donor will attack the cells that make up the tissues and organs of the host. This is called graft versus host disease. The favorite target organs of these killer T cells are the skin, the gastrointestinal tract, and the liver—and these attacks can be fatal. In general, the greater the difference between the MHC molecules of the donor and the patient, the stronger the graft versus host reaction. Moreover, even with a well-matched donor, graft versus host disease usually must be controlled by treating the patient with immunosuppressive drugs such as prednisone and cyclosporin for at least the first 100 days after transplantation.

Of course, one way to deal with graft versus host disease would be to eliminate all the killer T cells from the donor's marrow before the transplant takes place. This can be done, actually, but when bone marrow transplants are performed to try to cure leukemia, removing the killer T cells turns out to be a bad idea. When patients are treated with chemicals to try to kill off all their leukemic cells before a bone marrow transplant, sometimes the killing isn't complete. The result can be residual leukemic cells that eventually proliferate and spoil the treatment. For reasons that are not understood, killer T cells in the donor marrow don't like leukemic cells. Consequently, having killer T cells in the donated marrow is very important for "mopping up" any leftover leukemic cells, and for preventing a relapse of the disease. This is called the graft versus leukemia effect.

To decrease the chances of graft versus host disease, oncologists try to match MHC genes between donor and recipient. In fact, it is this endeavor which spawned the name "major histocompatibility." The word "histo" is

Greek for tissue, and when surgeons perform tissue (organ) transplants, they try to find donors whose MHC molecules are "compatible." The word "major" is also of importance here, because in addition to the major histocompatibility proteins, there are minor histocompatibility proteins that play a role (albeit a lesser role) in graft versus host disease (killer T cells don't like them either!). In fact, your best chance of receiving bone marrow that won't cause a major graft versus host reaction is to get that marrow from a sibling. Not only does a sibling have a higher likelihood than some unrelated person of having most of the same MHC molecules as you have, there is also a good chance that many of your sibling's minor histocompatibility proteins will be the same as yours.

Of course, one situation in which both major and minor histocompatibility molecules can be matched exactly is when the donor and the recipient are identical twins. In this sense, having an identical twin is great, because it gives you a source of compatible "spare parts." Otherwise, getting an exact MHC match is very difficult. Even the probability of having a sibling whose MHC genes match yours exactly is only about 25%. Clearly Mother Nature didn't foresee that we would be swapping bone marrow, or she wouldn't have made MHC genes so diverse.

Stem Cell Transplants

Because bone marrow transplants are the only way to cure certain types of leukemia (e.g., chronic myeloid leukemia), oncologists have been searching for ways to make this procedure less threatening—both to the patient and to the donor. In the early 1990s, there was a two-year-old boy from Virginia Beach, Virginia, who had chronic myeloid leukemia. Unfortunately, Michael had a rather rare combination of MHC genes, so it had been impossible to find a donor whose MHC molecules were sufficiently well matched to attempt a bone marrow transplant. At that time, however, Michael's mother was pregnant with a child who was to be Michael's little sister, Christina. Doctors determined that Christina's MHC profile was very similar to Michael's—similar enough to make Christina a suitable donor for a bone marrow transplant. Of course, this seemed like a moot point, for it would be unethical to use Christina as a bone marrow donor without her consent—and, after all, she hadn't even been born yet!

Then Dr. John Wagner and his associates decided to try a daring experiment. They knew that in an adult, most stem cells are found in the bone marrow. However during fetal development, stem cells are first found in the yolk sac that nourishes the fetus, and then in the fetal liver, and finally in the fetal bone marrow. Wagner reasoned that with all this moving around, there might be enough stem cells in Christina's circulating blood to do the trick for Michael. Wagner also knew that there is quite a bit of fetal blood in the umbilical cord—a piece of tissue that usually is discarded at birth. So when Christina was born, Wagner and his team drew about five tablespoons of blood from Christina's umbilical cord, and froze the white blood cells in this sample to use for Michael's transplant. Meanwhile, Michael was subjected to chemotherapy to try to kill all his leukemic cells—and then the white blood cells from Christina's umbilical cord were thawed and infused into Michael's blood stream.

The oncologists' first concern was that there might be too few stem cells in the umbilical cord blood to renew Michael's blood system. So you can imagine their excitement when they found that his blood system was recovering just fine after the transplant. Next they worried that there might be a problem with graft versus host disease, because Christina's MHC fingerprint was not an exact match to Michael's. However, it turns out that graft versus host disease is usually less of a problem with umbilical cord blood transplants than with bone marrow transplants—perhaps because killer T cells in umbilical cord blood are less "mature" than killer T cells in adult bone marrow. Finally, as with all treatments for leukemia, Michael faced the potential problem of residual disease, caused by leukemic cells that had escaped killing by the chemicals used to treat his leukemia. In fact, although no leukemic cells could be detected for the first two years after Michael's transplant, his leukemia did eventually return, with fatal consequences.

Sadly, the umbilical cord blood transplant did not cure Michael's leukemia, but this "experiment" helped open up a whole new area of transplant biology. Each year, about 100,000 umbilical cords are discarded in this country, and companies now have been set up to "bank" some of this umbilical cord blood for later use in transplants. For example, when my grandson, Andrew, was born, I arranged to have his umbilical cord blood frozen. Now, if he or another member of his family should need a "bone marrow" transplant at some time in the future, the stem cells in his cord blood will be available for them to use.

Even more recently, immunologists have found that, although blood from adults is not a rich source of stem cells, there are some stem cells in the adult circulation. In fact, treatment with certain drugs can cause many stem cells to leave the marrow, greatly increasing the number of stem cells in the blood. This procedure is called stem cell mobilization. So to avoid having a big needle inserted into their bones, donors can now have their stem cells mobilized and collected from their blood. Then the stem cells can be frozen for safekeeping, and at the proper moment, thawed and infused into the bloodstream of a leukemia patient. Recent evidence suggests that such "stem cell transplants" are at least as good as the traditional bone marrow transplants in curing leukemia.

In patients who are younger than about fifty years of age and who are in the chronic phase of chronic myeloid leukemia, the current treatment of choice is a bone marrow or stem cell transplant—if a matched donor can be found. Of these patients, roughly half will survive the transplantation procedure and be cured. For patients whose leukemias have already progressed to the acute stage, the five-year survival rate for individuals receiving a transplant is only about 6%. So having a bone marrow or stem cell transplant early in the course of the disease is important.

Gene-Specific Therapy

About a third of all patients with chronic myeloid leukemia are over fifty, and are not considered to be good candidates for these transplants: They have a high risk of not surviving the procedure. Because bone marrow transplants are risky, matched donors are difficult to find, and cures are not assured, there is adequate stimulus to look for new ways to treat chronic myeloid leukemia. The realization that the BCR-ABL protein is an "always-on" kinase enzyme led to a search for drugs that might specifically inhibit this enzyme and turn off the corrupted control system. Gleevec is the first drug of this type that has been approved for treating chronic myeloid leukemia. The story of how Gleevec was produced is quite interesting, because it gives insights into the process of drug discovery.

In 1990, researchers at the drug company Novartis engineered bacteria to produce the ABL protein. Their idea was to screen a large number of chemicals to see if any of them could inhibit the ability of this "synthetic" ABL kinase to add phosphates to other proteins. If such a chemical compound could be found, it might also inhibit the activity of the BCR-ABL oncoprotein in leukemic cells. Eventually, they identified one chemical which seemed to have some potential. However, this "lead compound" was not a very strong inhibitor of the ABL kinase, and it was not very specific: It inhibited several other kinase proteins that are found in normal cells.

Undaunted, the team at Novartis began making slight chemical modifications to the structure of the lead compound in an attempt to make it more potent and more specific. After two years of painstaking work, they succeeded in producing a modified compound that had the desired properties, and they called this chemical STI571. The reason for coining this name is that the BCR-ABL kinase sticks a phosphate onto an amino acid called tyrosine in the proteins it phosphorylates. Consequently, STI571 is a specific tyrosine kinase inhibitor. The number 571 gives you an idea of how many modifications of the lead compound they had to make to come up with one that worked!

The next step was to test whether STI571 could slow the growth of cells taken from patients with chronic myeloid leukemia. Happily, the compound passed this test. Not only did STI571 inhibit the proliferation of leukemic cells, but, importantly, the drug did not slow the growth of normal cells.

One more obstacle had to be overcome before STI571 was ready to try as a treatment for leukemia. During their tests of STI571 on leukemic cells growing in the laboratory, scientists noted that if the compound was washed off, the cells began proliferating again. This suggested that STI571 would need to be administered more or less continuously to keep leukemic cells under control. Unfortunately, the original formulation of STI571 was intended to be administered to patients by injection, and the people at Novartis reasoned that to be practical, the drug would have to be given as a pill. So they went back to the drawing board, and made further modifications to the structure of STI571 to try to produce a drug that would survive the insults of the digestive system. Fortunately, they were successful, and the result was the pill form of STI571 known as Gleevec.

After successful testing in animals, trials began in 1998 to determine the usefulness of Gleevec in treating humans. These and subsequent tests were so successful that Gleevec was approved by the FDA for the treatment of chronic myeloid leukemia. What makes Gleevec so special is that it is one of very few mechanism-based cancer drugs—drugs designed to repair defective cellular control systems. It is hoped that as more is learned about how control systems are corrupted in cancer cells, more drugs like Gleevec will be discovered that can "make cancer cells well again."

SELECTING TREATMENTS BASED ON MUTATIONS

Although the outlook is fairly bleak for most individuals who suffer from the two leukemias we have discussed, I don't want to end on a negative note. Indeed, one of the greatest successes in the treatment of cancer has been with children who have acute lymphocytic leukemia. This is the number one cancer that affects children, and by using chemotherapy or bone marrow transplants, roughly 75% of these kids can now be cured. This is to be compared with a cure rate of only about 3% for this same disease thirty years ago. Two reasons for this dramatic increase are more effective chemotherapeutic drugs and improvements in bone marrow and stem cell transplantation protocols. However, equally important has been the realization that all acute lymphoblastic leukemias are not the same, and that the course of the disease depends in a large part on the particular mutations that are present in the patient's leukemic cells. Moreover, a knowledge of these mutations can be used not only to predict disease outcomes, but also to tailor the treatment to the genetic profile of the patient. For example, children with acute lymphoblastic leukemia who have translocations that involve the AML1 gene usually can be treated successfully without employing the most toxic (and therefore life-threatening) forms of chemotherapy. On the other hand, children whose leukemic cells contain the Philadelphia chromosome must be treated with much more aggressive chemotherapy or bone marrow transplants. Milder treatments usually are ineffective in patients with this genetic profile. Hopefully, when more is known about the biological basis of other types of leukemia, treatment protocols based on genetic alterations can be devised for those cancers as well.

TRANSLOCATIONS AND LEUKEMIA

In this lecture we have focused on two types of leukemia that frequently involve translocation mutations. However, you shouldn't get the idea that the mutations which cause leukemia always are due to translocations. For example, in about 3% of the cases of acute myeloid leukemia, the change of a single letter in the recipe for the AML1 protein (a "point" mutation) is implicated in helping cause the disease. One reason translocations appear to be so frequent in leukemias is that translocations are relatively easy to spot. Also, early leukemia research focused on this type of mutation because of the clear association between exposure to ionizing radiation and the occurrence of leukemia. However, as progress is made in technologies that allow detection of more subtle mutations, translocations will prove to be just one, very obvious type of mutation that can lead to leukemia.

WHY ARE SOME CANCERS MORE COMMON THAN OTHERS?

You may be wondering: If fewer control systems must be disrupted to create a leukemic cell than to produce a cancer cell which must metastasize to do its evil (e.g., a breast cancer cell), why isn't leukemia among the most "popular" cancers? After all, only about 3% of all cancer deaths in our country are due to leukemia. This is actually part of a broader and equally interesting question: What determines the frequency of occurrence of various cancers? Indeed, the human body is composed of roughly 200 different kinds of cells, yet the "common" cancers arise in only about a dozen of these cell types. The answer to this important question is not known for sure, but there are at least two possible reasons for this cell-type selection.

Cancer occurs when genes that make up growth-promoting and safeguard systems are mutated, and certain cell types are more likely to suffer mutations than are others. In general, cell types that proliferate rapidly are more susceptible to cancer-causing mutations because they have less time to repair damage to their DNA. So for example, the epithelial cells that line the colon proliferate almost continuously, and these cells are a popular target for cancer. In contrast, cells in the heart, which proliferate little if at all in an adult, rarely become cancerous.

Not only do mutations tend to occur more frequently in rapidly proliferating cells, but certain cells in the body are more "in harm's way" than others when it comes to mutations. For example, skin cells are in just the right place to be bombarded by mutation-causing UV radiation. In contrast, brain cells live in an environment that is usually protected from agents that cause mutations. Consequently, we would predict that brain cancer would be rare when compared with skin cancer. And this is certainly true.

So an important factor that influences whether a certain cell type will be "cancer prone" is the likelihood that those cells will suffer mutations. But there is a second factor that is probably equally important in determining which cell types become cancerous: Different cell types can have different control systems.

Except for egg and sperm cells (which are special), all the cells in your body have the same genes. What makes colon cells different from brain cells is that although all the same recipes are there, the frequency with which a given recipe is used to make the protein it specifies is different in different cell types. It's as if 200 different chefs have the same cookbook, but each one chooses to use only selected recipes in his cooking.

One result of this selective use of recipes is that the control systems and the proteins that make up these systems can differ from cell type to cell type. For example, erythropoietin is a growth factor that selectively triggers the proliferation of red blood cells. In contrast, erythropoietin has no effect on the proliferation of other cells in the body—because they don't have this particular control system. Because the probability that mutations will disrupt the function of a protein is different for every protein, and because different control systems are composed of different protein elements, some control systems are easier to corrupt than are others. In addition, some cell types have more "backup" safeguard systems than others. The bottom line is that the chance that a given cell type will become cancerous depends not only on how likely that cell type is to suffer mutations, but also on which control systems must be corrupted in that particular cell type to cause cancer.

THOUGHT QUESTIONS

1. What is the underlying cause of leukemia?
2. Explain what a translocation is, and how it can cause a mutation.
3. What types of control systems must be corrupted for a blood cell to become cancerous?
4. Why is it probably "easier" to become a blood cell cancer than a solid tumor (e.g., a breast cancer)? Could this explain why young people frequently get blood cell cancers?
5. What is an oncogene? Give an example.
6. What is a proto-oncogene? Give an example.

A SUMMARY OF CANCER CONCEPTS

The aim of these lectures is to use various types of cancer to illustrate principles that are important for understanding cancer in general. Here is a table in which the concepts discussed in this lecture are listed.

Table of Concepts for Lecture 2

Concept	Example
Cancer-associated mutations can result from translocations.	AML-ETO in acute myeloid leukemia BCR-ABL in chronic myeloid leukemia
Many cancers probably arise due to mutations that accumulate in immortal stem cells.	Acute myeloid leukemia
Environmental factors can increase the risk of certain cancers.	Ionizing radiation and leukemia
Multiple control systems must be corrupted to produce a cancer cell.	Acute myeloid leukemia
Oncoproteins can short-circuit growth factor pathways.	BCR-ABL protein
Oncoproteins can block cell maturation.	AML-ETO protein
Bone marrow or stem cell transplants can be used to treat cancer.	Leukemia
Treatments can be tailored to mutation profile.	Chronic myeloid leukemia Acute lymphocytic leukemia
Drugs can heal corrupted control systems.	Gleevec for chronic myeloid leukemia

Lymphoma

REVIEW

All blood cells are born in the bone marrow from self-renewing stem cells. After a period of proliferation, most of these cells mature inside the marrow. If, however, multiple control systems are corrupted so that one of these "baby" cells proliferates within the bone marrow but fails to "grow up," the result is leukemia. In this disease, immature (and therefore, useless) cells fill up the marrow, making it impossible for normal cells to mature. Left untreated, a patient with leukemia dies because he does not have sufficient red blood cells to carry oxygen or sufficient white blood cells to repair damaged blood vessels and protect him from disease.

Many of the mutations that have been identified in leukemic cells are translocations in which chromosomes have been broken and rejoined incorrectly. In acute myeloid leukemia, such a translocation creates a fusion gene, AML-ETO, which specifies a protein that interferes with the maturation program of the cell. However, this mutation alone is not sufficient to turn a blood cell into a leukemic cell. Other mutations are required. Indeed, there is no human cancer which has been shown to result from a single mutation—multiple mutations are needed to disrupt the systems that control cell growth.

A BCR-ABL fusion gene (the Philadelphia chromosome) is involved in essentially all cases of chronic myeloid leukemia. In contrast to the AML-ETO fusion, which blocks maturation, the BCR-ABL mutation short-circuits a growth factor pathway. This pathway is made up of proteins that relay the "grow" signal all the way from the surface of the cell, where growth factors bind to their receptors, to the nucleus of the cell, where the genes involved in making the cell proliferate are located. By short-circuiting this pathway, the BCL-ABL protein causes the cell to proliferate even though there are no growth factors outside the cell signaling that proliferation is required.

The BCR-ABL mutant gene is an excellent example of an oncogene—a rogue gene that specifies a protein (an "oncoprotein") which, by its action, can contribute to a cell becoming a cancer cell. The ABL gene, which causes the production of a perfectly normal protein, is called a proto-oncogene: a gene which, if mutated, can become an oncogene. In all cancers studied so far, oncogenes are dominant: Only one bad copy of a gene (e.g., one BCR-ABL fusion gene) is required to disrupt growth control. So having a second, good copy of the ABL gene won't rescue a cell from the effects of the BCR-ABL oncogene.

Although cells with the BCR-ABL translocation proliferate inappropriately and disrupt the balanced production of various types of blood cells, they are mature, fully functional cells. Consequently, additional mutations must occur if cells with the BCR-ABL translocation are to acquire both of the signature characteristics of leukemic cells: uncontrolled proliferation and a block in maturation.

One treatment that has been successful in curing chronic myeloid leukemia (and several other leukemias) is a bone marrow or stem cell transplant. In this procedure, stem cells from a donor are used to restore the blood system of a leukemia patient after chemotherapy has destroyed all his cancerous blood cells. A new drug called Gleevec is also being used to treat some cases of chronic myeloid leukemia. This drug acts by reversing the effects of the BCR-ABL oncoprotein. Gleevec is one of the first examples of a drug which can restore to health a control system that has been corrupted by the action of an oncoprotein.

LYMPHOMA

Most types of blood cells remain within the bone marrow as they finish maturing, eventually becoming fully functional "adult" cells. However, one class of blood cells, the lymphocytes, leave the bone marrow before their maturation program has been completed. This means that lymphocytes can be arrested in their development and become cancer cells either before or after they leave the marrow. To fully understand how lymphocytes become cancerous, we need to take a minute to discuss the place where lymphocytes finish growing up: the lymphatic system.

In your home, you have two plumbing systems. The first supplies the water that comes out of your faucets. This is a pressurized system with the pressure being provided by a pump. In addition, you have another plumbing system that includes the drains in your sinks, showers, and toilets. This second system is not under pressure—the water just flows down the drains and out into the sewer. The two systems are connected in the sense that eventually the waste water is recycled and used again.

The plumbing in a human is very similar. We have a pressurized system (the cardiovascular system) in which blood is pumped around the body by the heart. Everybody knows about this one. But we also have another plumbing system—the lymphatic system—which is not under pressure, and which drains the fluid (called lymph) that leaks out of our blood vessels into our tissues. Without this system, our tissues would fill up with fluid and we'd all look like the Pillsbury Doughboy. Fortunately, this doesn't happen, because lymph is collected from our tissues and is transported by muscular contraction through a series of one-way valves to the upper torso, where it is recycled back into the blood via the right or left subclavian veins.

So one function of the lymphatic system is to drain and recycle fluid from our tissues. However, this plumbing system serves another important function: The lymphatic system plays an essential role in the immune system's response to infection. From this diagram, you can see that as the lymph winds its way back to empty into the blood, it passes through a series of way stations, the lymph nodes.

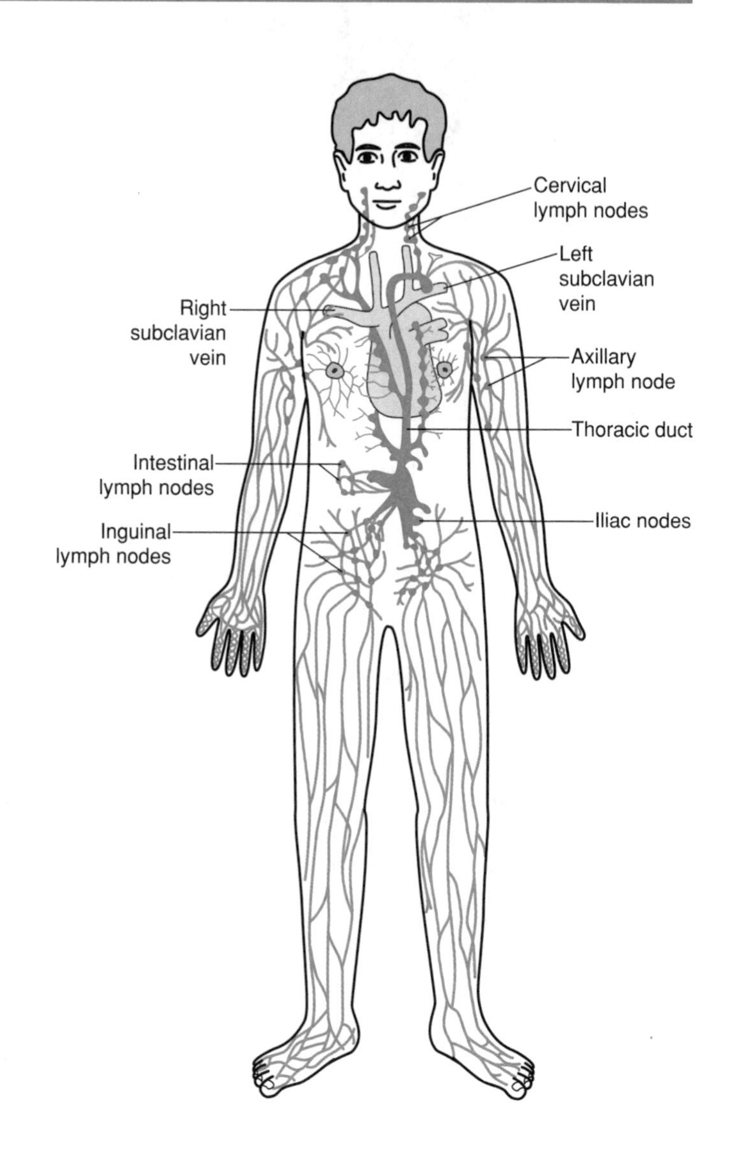

In your body there are thousands of lymph nodes, ranging in size from very small to almost as big as a Brussels sprout (only a few are shown in this diagram). When bacteria and viruses invade our tissues (e.g., because of a wound), they are carried by the lymph to nearby nodes. Meanwhile, immune system cells called T and B lymphocytes circulate from node to node, looking for invaders. They are named "lymphocytes" because they spend so much of their time hanging out in the lymphatic system.

In a sense, lymph nodes function as "singles' bars"—places where immune system cells and invaders meet. In fact, it is only by bringing lymphocytes and invaders together in the small volume of a lymph node that the immune system can be activated to repel an invasion. This requirement for a "close encounter" between invaders and immune system cells arises because only about one lymphocyte in a million will recognize a particular invader. Consequently, it is important that each invader be "interviewed" by a lot of lymphocytes.

To get a feeling for how and where lymphocytes mature, let's follow a B cell as it grows up. When B lymphocytes mature in the bone marrow, they begin to express protein molecules on their surface called B cell receptors. These receptors are the "antennae" that B cells will eventually use to identify invaders. The "baby" B cells are loosely tethered within the bone marrow by "sticky" molecules on their surface that adhere to the structures that organize the marrow. This tethering helps retain B lymphocytes within the bone marrow until they have completed the initial stages of their maturation program.

After a B cell has matured to the point where its receptors are posted on its surface, the cell is ready to go looking for the invader that its particular receptors will recognize. At this stage, these cells can be thought of as "adolescent" B cells. During the transition from babyhood to adolescence, important changes take place. First, the tethering molecules on the B cell surface are removed, freeing the cell to enter the blood. In addition, because the B lymphocyte is most likely to find an invader in one of the many lymph nodes, the B cell is issued a "passport" that makes it possible for the cell to leave the blood and enter the lymphatic system. This passport consists of special molecules that are displayed on the surface of the B cell and which interact with "doorkeeper" molecules on lymph nodes. Guided by these molecules, adolescent B cells circulate through the blood, enter the lymphatic system, and visit many lymph nodes before they are dumped back into the blood system along with the lymph.

This circulation from blood to lymph to blood doesn't continue for very long: The B cell has only a week or two to search through the lymphatic system, looking for the invader to which its receptors will bind. If it finds that invader, the B lymphocyte can mature to the "adult" stage and become an antibody factory. Antibodies are protein molecules that can bind to invaders and tag them for destruction. The production of large numbers of antibodies is one of the main ways the immune system defends us against infections by viruses, bacteria, and parasites.

If a B cell does not find the invader its receptors recognize during its travels through the lymphatic system, that B cell is programmed to die. Because humans are generally under attack by only a few different invaders at any given time, most B cells never find their targets. Indeed, billions of B cells die each day without meeting up with the invader for which they are "fated."

For our discussion here, it is important to note that baby and adolescent B cells have different traffic patterns. The babies are found in the bone marrow and the blood, but not in lymph nodes. In contrast, adolescent B cells hang out mainly in the lymphatic system. So although all blood cell cancers arise when corrupted control systems cause a blood cell to proliferate without maturing, this block in maturation can happen at different stages in the cell's development. Indeed, blood cell cancers are classified according to the stage at which the maturation process is blocked.

Cells which stop maturing very early while they are still "babies" in the bone marrow are called leukemic cells. Because all blood cells go through their early stages of maturation in the bone marrow, any type of blood cell (even a red blood cell or a lymphocyte) can become a leukemic cell. These baby cells do not have passports that allow them to enter lymph nodes, so they accumulate in the bone marrow and in the blood of a leukemia patient, but not in his lymph nodes.

In contrast, when lymphocytes become "adolescents," they are issued passports to lymph nodes. If cellular control systems are corrupted, blocking their further maturation, adolescent lymphocytes can accumulate in lymph nodes, causing lymphoma.

Finally, if a B cell reaches "adulthood" and becomes an antibody-producing cell (sometimes called a plasma cell), a cellular control system normally triggers these cells to die after about five days. However, if this "life-span control system" malfunctions, the adult B cell can become "immortal," resulting in a cell that produces antibodies, but doesn't take that final step in its maturation—death. Such cancerous cells are called myelomas.

TYPES OF LYMPHOMA

So lymphoma results when adolescent lymphocytes refuse to grow up. Lymphomas are generally divided into two categories: Hodgkin's lymphoma (also called Hodgkin's disease) and non-Hodgkin's lymphomas (how original!). Hodgkin's lymphoma usually affects young adults between the ages of about 15 and 35. The most common early symptom associated with this disease is a painless swelling of lymph nodes in the upper part of the body. The genes you inherit clearly can effect your risk of developing Hodgkin's lymphoma. For example, the same-sex sibling of someone who develops this lymphoma has about a 10-fold higher risk of getting the disease than does an unrelated person. So far, however, no mutations have been definitively identified which, if inherited, increase one's chances of getting Hodgkin's lymphoma. The good news is that Hodgkin's disease is one of the most curable of all cancers: Roughly 80% of Hodgkin's disease patients can be cured by chemotherapy or chemotherapy plus radiation therapy. So if you get to choose your lymphoma, choose this one. The bad news is that Hodgkin's lymphomas represent only about 15% of all lymphoma cases in the United States.

The non-Hodgkin's category, which includes "all the others," represents a rather diverse collection of diseases. Indeed, there are at least ten different types of non-Hodgkin's lymphomas. For our discussion, I have chosen "follicular" lymphoma as our model non-Hodgkin's lymphoma, because follicular lymphoma has been studied extensively, and it can teach us a lot about lymphomas in general.

FOLLICULAR LYMPHOMA

Lymphomas are also categorized according to the type of lymphocyte (either B cell or T cell) the cancer cell was destined to become when things went wrong. About 80% of all lymphomas are B cell lymphomas—lymphomas that arise when control systems within a B lymphocyte are corrupted. The remainder are T cell lymphomas. A follicular lymphoma is a B cell lymphoma, and its name derives from the fact that the B lymphocyte which gives rise to this cancer was in a region of a lymph node called the follicle when its maturation program stalled out. It is in the follicle of a lymph node that a B cell's receptors are "fine tuned" so that they will bind more avidly to invading pathogens. The way this works is pretty amazing.

In the follicle of a lymph node, B cells are triggered to mutate at an extremely high rate. Fortunately, this high rate of mutation usually is confined to the genes that specify the proteins which make up the B cell's receptors. After these mutations have occurred, the receptors are tested. Those B cells whose receptors have mutated to bind more tightly to their favorite pathogen are triggered to proliferate more rapidly than B cells whose receptors bind less tightly. As a result of this elegant mutation and selection scheme, the tight binders win out, yielding a pool of B cells whose receptors can bind very tightly to an invader.

This fine tuning of B cell receptors is great for the immune system, because the mutated B cells end up working better to protect us from invaders. However, this immune improvement is not without risk, because sometimes when the B cell mutates, mistakes are made that can help turn a normal B cell into a lymphoma. Although the underlying mechanism is not completely understood, the most common "mistake" is one that results in a specific type of translocation—one in which part of the gene for one of the B cell's receptor proteins is joined to part of some other cellular gene. Importantly, the piece of the receptor protein gene that is translocated contains the "control region"—the part of the recipe that specifies how many copies of the protein should be made. Because each B cell needs to make many receptor proteins, the fusion protein that results from the translocation is also produced in great abundance. And that's the problem.

In about 80% of the cases of follicular lymphoma, a translocation occurs in which part of the gene for the B cell receptor on chromosome fourteen is joined to part of the gene for a protein called bcl-2 located on chromosome eighteen. As a result of this 18;14 translocation, way too much of the bcl-2 protein is made.

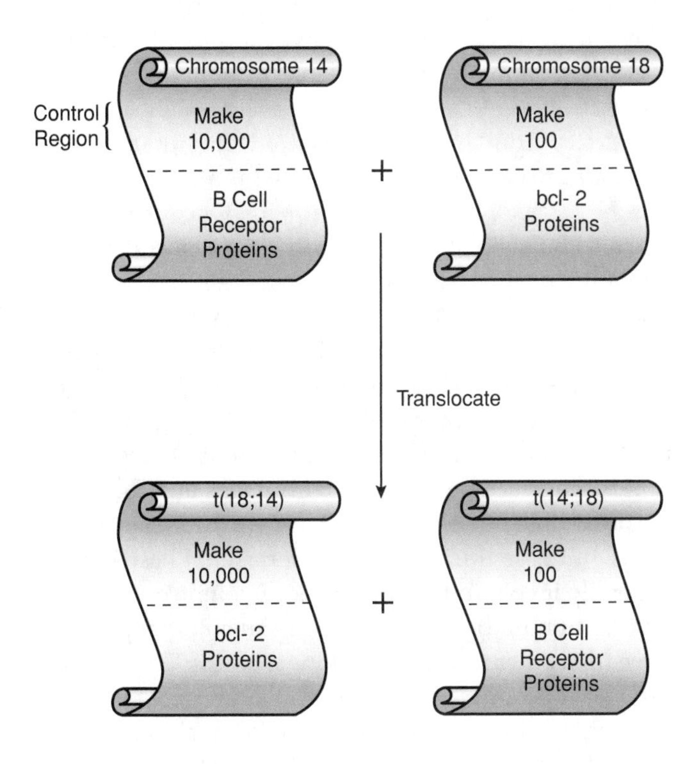

Now it turns out that this bcl-2 protein is a "survival" protein—a protein that keeps B cells from dying. Normally, after a B lymphocyte has lived its usual life span, the bcl-2 protein either stops being produced or its survival effects are negated by other proteins that begin to be made as the cell ages. Indeed, B cells are programmed to die within a couple of weeks if they don't find the invader their receptors recognize. And even if they do find their "partners," most B cells mature into antibody-producing cells which are scheduled to die after about five days. So part of a B cell's normal maturation program is death. This makes sense. You don't want a whole bunch of old B cells hanging around that recognize last month's invader. You want "room" in the B cell pool for new B cells that can recognize the microbes that are causing today's infection.

The problem that arises when the bcl-2 protein is overproduced because of the 18;14 translocation is that B cells with this mutation refuse to die. They become "immortal." Biologists have now produced mice whose B cells have the 18;14 translocation. Although their B lymphocytes do not proliferate any more rapidly than do normal B cells, the lymph nodes of these mice eventually fill up with B cells that never fully mature—they just won't die.

So the normal function of the bcl-2 protein is to promote cell growth by keeping a cell alive. However, when the bcl-2 protein is overproduced because of the 18;14 translocation, this growth-control system is corrupted, the B cell does not die when its useful life is over, and the result is follicular lymphoma.

Because the fusion protein specified by the 18;14 translocation promotes inappropriate cell growth, the translocated bcl-2 gene is considered to be an oncogene. Likewise, the normal bcl-2 gene fits the description of a proto-oncogene—a gene that, when mutated, specifies a protein which, by its action, can contribute to a cell becoming a cancer cell.

It is important to note that some perfectly normal humans have B cells with 18;14 translocations, so this mutation is certainly not the whole story. The current view is that although the 18;14 translocation can play an important role in producing a follicular lymphoma, additional mutations are required. Indeed, it is likely that the overexpression of the bcl-2 protein artificially lengthens the life of B cells in the lymph node follicles. Then, on rare occasions, mutations may occur which corrupt other control systems in one of these long-lived cells, producing a full-blown lymphoma. So in a sense, the bcl-2 mutation may "set the stage" for other mutations.

Having escaped death, large numbers of immortal lymphoma cells accumulate in the lymph node in which they originated. So it is no wonder that patients with follicular lymphoma usually first visit their doctor complaining of painless, swollen lymph nodes—nodes that are chock full of lymphoma cells. Eventually these cells spill out of this node into the lymph, and travel to the next node in the line, where they can take up residence and continue to proliferate. Because lymph eventually is recycled back into the blood stream, lymphoma cells can also be found in the blood, the liver, the spleen, and the bone marrow.

Follicular lymphoma usually begins as a slowly progressing (indolent) disease, because initially the lymphoma cells are proliferating relatively slowly. However, within about ten years of diagnosis, roughly half of all patients will progress from this "low grade" lymphoma to a "high grade" lymphoma in which their lymphoma cells proliferate much more rapidly. This sequence of events is very similar to the progression from the chronic to the acute stage of chronic myeloid leukemia, and it is presumed that this change from low to high grade results when additional mutations disrupt growth-control systems within the cell.

There are thousands of lymph nodes in the human body, and many of these are in close proximity to vital organs. Eventually, lymphoma cells growing out of control can crowd out these organs, disrupt their function, and cause the death of the patient. Of course, if we had no lymph nodes, nobody would get lymphoma. However, we depend heavily on the lymphocytes which mature in these nodes to protect us from infectious diseases (e.g., viral infections). Indeed, without lymph nodes, none of us would live long enough to get cancer of any kind! So again we see that the possibility of getting cancer—in this case, lymphoma—is an unavoidable consequence of being human.

DRUG DISCOVERY AND CLINICAL TRIALS

The next logical question is, "How do you treat follicular lymphoma?" But before we discuss therapies for this particular cancer, I want to take a moment to outline the process by which new treatments are discovered and tested. As you will see, developing a new, anti-cancer drug is a long, complicated, and costly procedure.

Most anti-cancer treatments are based on discoveries made using cancer cells growing in petri dishes in the laboratory. For example, a pharmaceutical company may screen thousands of compounds to see if any of them stops cancer cells from growing, yet allows normal cells to proliferate. A compound with this property might be effective as an anti-cancer drug. However, what works on cells growing in the laboratory doesn't always work in the "real world." For example, a potential anti-cancer drug might be so toxic that it would kill a human before it had any real effect on his cancer. So the next step in drug discovery usually is to test the compound on animals. Mice are quite popular for this because they are cheap to purchase and maintain, and their responses to drugs are frequently (but not always!) similar to those of a human. A typical experiment would be to inject human cancer cells under the skin of a mouse, allow these cells to proliferate to form a tumor mass there, and then treat the mouse with various doses of a potential anti-cancer drug. If this experimental therapy caused the donated tumor to stop growing at doses that didn't make the mouse too sick, the compound would be considered promising. This type of test is usually referred to as a preclinical trial—probably to keep the animal rights people happy.

If a therapy is shown to be safe and effective in animals, the next step is to carry out a Phase I human trial. This is an important test, because a therapy that works in animals may be ineffective or may have serious side effects in humans. For a Phase I trial, a small number of volunteers (generally about a dozen) is recruited. These are usually people who have cancer and who have tried the best treatments available without success. The goal of a Phase I trial is to determine dosages of an experimental drug that can be tolerated without unacceptable side effects. In essence, a Phase I trial is intended to test the safety of a potential treatment. Of course, everyone also hopes that the drug will have a positive effect on the volunteer's cancer, but the success or failure of a Phase I trial does not depend on this point.

Based on experiments in animals, scientists usually have some idea of the dose of a drug which is required for an anti-cancer effect (e.g., to stop growth of a tumor). If the results of a Phase I trial indicate that doses in this range can be tolerated by humans and that there are no terrible side effects (i.e., that the drug is safe for humans), a Phase II trial can be undertaken. Phase II trials generally involve a larger number of volunteers (usually several hundred). Again, these are mainly people who have tried other treatments without success. This can be a problem, of course, because these folks are generally very sick, and their cancers are usually quite far advanced—so it is really asking a lot for an experimental therapy to help them. Nevertheless, the goal of a Phase II trial is to further evaluate safety, and also to look for some anti-cancer effect. Because it involves a fairly large group of people, it is likely that if the drug has any value at all, some of these people will see an improvement in their condition.

If a Phase II trial indicates that a drug can be tolerated at doses that show some anti-cancer activity, a Phase III trial is the next step. The goal of a Phase III trial is to evaluate the effectiveness of a drug, and to test for infrequently occurring side effects that might have been missed in a smaller, Phase II trial. These trials involve a much larger number of volunteers (usually more than 1,000) whose cancers are at an earlier stage. Phase III trials are "controlled" studies in which some of the volunteers may be given the best available therapy, while some are given the experimental drug. However, this design frequently raises an important ethical question: Is it proper to offer a patient an experimental drug which may turn out to be of little value, when a treatment already exists that has been shown to be helpful? It's a tough question. One way to deal with this dilemma is to structure the Phase III trial in such a way that one group is given the best available treatment, while the other group is given that same treatment plus the experimental drug. However, this compromise design has the drawback that the combination of the two treatments may mask the value of the experimental drug.

If a Phase III trial shows that the drug is effective and has tolerable side effects, it can be licensed by the Food and Drug Administration for general use. It usually takes several years for a drug to be tested in Phase I through III trials, and it generally costs upwards of $500 million for a pharmaceutical company to bring a new anti-cancer drug to market! It is small wonder that these drugs are so expensive.

TREATMENTS FOR FOLLICULAR LYMPHOMA

Presently, it is not possible to cure follicular lymphoma. However, because follicular lymphoma starts out as a low-grade, indolent cancer, oncologists currently view this type of lymphoma as a chronic disease which can be "managed" using a "watch and treat" strategy.

Standard Chemotherapy

Patients with follicular lymphoma can be treated with standard chemotherapeutic drugs (e.g., cyclophosphamide) which damage cellular DNA, and can kill lymphoma cells. However, there is a problem with this approach. These drugs work best on cancer cells that are proliferating rapidly—and low-grade follicular lymphoma cells are not. Unfortunately, by the time the disease has progressed to a higher grade in which the cells are proliferating rapidly, enough other mutations will have occurred to make treatments with DNA-damaging agents problematic. It is ironic that cancers that proliferate slowly are generally harder to treat than those which proliferate rapidly.

Stem Cell Transplants Plus Chemotherapy

Because there is so little difference in the proliferation rates of low-grade lymphoma cells and normal lymphocytes, doses of chemotherapeutic drugs which might kill the lymphoma cells will also destroy many normal cells. To try to get around this problem, oncologists have devised a strategy that combines standard chemotherapy with a stem cell transplant. Here's how it works.

A patient with follicular lymphoma is first given drugs which increase the number of stem cells that are circulating in his blood. These "mobilized" stem cells are then removed from the blood and stored. Next the individual is treated with high-dose chemotherapy. The idea here is that by removing the patient's stem cells before chemotherapy begins, very high doses of these drugs can be administered to try to destroy most of the lymphoma cells. When the chemotherapy is finished, the stored stem cells (which would have been killed by the chemotherapy had they remained in the body) can then be returned to the patient's blood stream (and eventually his bone marrow) to rebuild his blood system. This approach has been used to achieve long-lasting remissions for patients with follicular lymphoma.

Using Antibodies to Treat Cancer

Very early on, immunologists envisioned using antibodies as "magic bullets" to destroy tumor cells. For example, if antibodies could be made which could bind to a specific protein on a cancer cell, oncologists might be able to inject these antibodies into a patient and destroy his cancer. The first problem with this approach was the difficulty in making a lot of these special antibodies. One possibility would be to vaccinate a mouse with the "cancer protein," and isolate from the blood of the mouse those particular antibody-producing B cells (plasma B cells) which make antibodies that bind to the cancer protein. Then one might grow up a lot of these plasma B cells in the lab, and use them as "antibody factories" to make a ton of anti-cancer antibodies. Unfortunately, there's a serious flaw in this approach: Plasma B cells only live for a few days. So to allow time to identify a particular B cell that makes the desired antibody, and to grow up a lot of them, a procedure would have to be devised that would extend the lives of plasma B cells—and that's exactly what two scientists named Kohler and Milstein were able to do.

Kohler and Milstein recognized that many cancer cells are "immortal," and can be grown almost indefinitely in the lab. So to begin their experiments, they isolated from a mouse, immortal, cancerous B cells (myeloma cells) that did not produce antibodies (an unusual type of myeloma). Next they vaccinated another mouse with a chicken protein (the test protein they wanted their antibodies to recognize), harvested B cells from this vaccinated mouse, and fused these B cells with the immortal myeloma cells to make hybrid cells. Their hope was that the hybrids, which were part cancer cell and part plasma B cell, would have the best qualities of each parent: that they would inherit the ability to grow indefinitely in the lab from their myeloma parent, and that, like their plasma B cell parent, they would produce antibodies that could bind to the chicken protein. If this strategy worked, Kohler and Milstein would have produced a "hybridoma" cell that could proliferate indefinitely in the lab, and which could pump out huge quantities of antibodies that would recognize their test protein. This plan seemed almost certain to fail. After all, who would have bet that these guys could "mate" two kinds of mouse cells and produce an antibody factory. But succeed they did, and for their discovery, Kohler and Milstein were awarded Nobel Prizes.

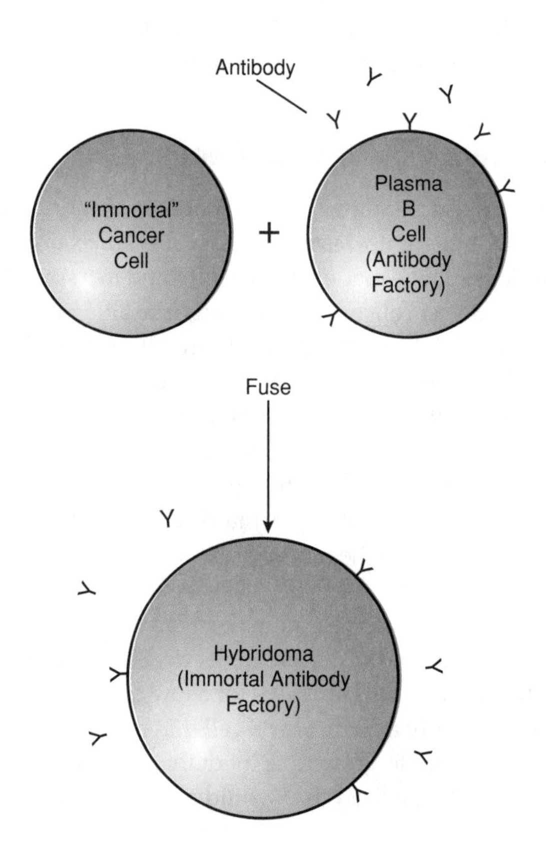

The reason this technological breakthrough is so important is that hybridoma cells, which are part cancer cell and part plasma B cell, can produce large amounts of almost any antibody you want. Antibodies made using this technology are called mouse "monoclonal" antibodies, because the mouse hybridoma cells make antibodies that are all the same, and which therefore bind to the same invader or protein. In a sense, Kohler and Milstein discovered a way to "clone" antibodies.

Rituxan

Recently, a new drug called Rituxan was introduced to treat follicular lymphoma. Rituxan is a monoclonal antibody that is made using a modification of the technique pioneered by Kohler and Milstein. I'm sure the irony is not lost on you that this drug, which is used to treat blood cell cancer, was produced using a cancerous blood cell! Anyway, Rituxan was the first monoclonal antibody approved by the FDA for the treatment of cancer, and over 100,000 patients have now been treated with this drug. In contrast to standard chemotherapy, which is quite nonspecific, Rituxan preferentially targets lymphoma cells. This monoclonal antibody binds to a protein called CD20 that is present on the surface of the follicular lymphoma cells of most patients with this disease. Importantly, CD20 is not found on the surface of blood stem cells or on adult B cells that have already matured to the antibody-producing stage. This means, at least in principle, that Rituxan can bind to lymphoma cells, targeting them for destruction, without damaging the stem cells whose function is required to restock the blood system. Moreover, Rituxan treatments should spare those "good" B cells that have not been blocked in their maturation program, and which are producing antibodies that can protect against infections.

Although treatment with Rituxan has been shown to selectively destroy follicular lymphoma cells, it is not clear how this killing takes place. Nevertheless, Rituxan causes remissions in about 50% of all cases of follicular lymphoma. Why the other 50% don't respond to the monoclonal antibody treatment is not known.

Antibodies That Recognize B Cell Receptors

The "Holy Grail" of monoclonal antibody treatments is to produce an antibody which recognizes a protein that is present on the surface of cancer cells, but which is not found on the surface of normal cells. This is a tall order.

Even Rituxan binds to some normal blood cells that display the CD20 protein.

One target on a follicular lymphoma cell that is truly unique is the B cell receptor. This receptor is the antenna that the B cell uses to identify invaders, and because of the way the proteins for B cell receptors are produced, every B cell is born to have a different B cell receptor. Consequently, the original lymphoma cell and its descendants will have a unique B cell receptor molecule on their surfaces—and this should be an ideal target for anti-cancer antibodies.

Monoclonal antibodies have now been made that recognize the B cell receptors on individual patients' lymphoma cells. In one study, many of the thirteen patients treated with these antibodies showed some improvements (one complete remission and eight partial remissions). Unfortunately, the treatment eventually became ineffective in most patients. The reason is that like most other cancer cells, lymphoma cells mutate rapidly. Consequently, in the huge collection of lymphoma cells in a patient's body, there are generally cells which have mutations that can render the B cell's receptor unrecognizable to the monoclonal antibody. This underscores a general problem encountered with most cancer treatments: Cancer cells usually mutate so rapidly that they can evade attempts to destroy them.

This type of antibody therapy also has the drawback that a different monoclonal antibody must be derived for each patient—because the receptor proteins on each patient's lymphoma cells are unique. Consequently, monoclonal antibodies must be tailor-made for each individual patient, making these "personalized" treatments very costly.

THOUGHT QUESTIONS

1. What is the underlying cause of leukemia and lymphoma?
2. What differentiates a leukemia cell from a lymphoma cell?
3. What are the hallmarks of blood cell cancers?
4. Why is it usually harder to treat cancers which grow slowly than cancers which grow rapidly?
5. Why are monoclonal antibodies potentially so useful as a cancer therapeutic?
6. Why do many cancer treatments fail to destroy all the cancer cells in a patient's body?

Table of Concepts for Lecture 3

Concept	Example
Oncoproteins can block normal cell death.	bcl-2 in follicular lymphoma
Bone marrow or stem cell transplants can be used to treat cancer.	Follicular lymphoma
Monoclonal antibodies can be used to treat cancer.	Rituxan for follicular lymphoma
Cancers frequently contain mutant cells that can resist anti-cancer treatments.	Monoclonal antibody-resistant lymphomas

Breast and Prostate Cancer

REVIEW

Unlike most other blood cells, B cells and T cells (the "lymphocytes") complete their maturation programs outside the bone marrow in the lymphatic system. Consequently, lymphocytes can either become leukemic cells—if their maturation is blocked early, while they are "babies" in the bone marrow—or become lymphoma cells—if their maturation program is halted later, after they have left the marrow and entered the lymphatic system as "adolescents." These cancerous lymphocytes proliferate in the lymph nodes, filling them to overflowing. Eventually, the bloated lymph nodes can crowd out vital organs, leading to organ malfunction and death. So blood cell cancers (e.g., leukemia and lymphoma) result when multiple control systems are corrupted that allow cells to proliferate without maturing. If the block in maturation occurs while the cells are still in the bone marrow, the result is leukemia. If the maturation program is halted while a lymphocyte is growing up in a lymph node, the consequence is lymphoma.

Follicular lymphoma results when a B lymphocyte ceases maturing while in the region of a lymph node called the follicle. It is in the follicle that a B cell's receptors are "fine tuned," and it is believed that mistakes made during fine tuning cause the mutation or mutations that eventually lead to follicular lymphoma. B cells usually live a very short time as "adults," being programmed to die after they have done their thing, so death can be viewed as the final stage of B cell maturation. In most cases of follicular lymphoma, a translocation occurs in which the gene for the bcl-2 "survival protein" is mutated so that the recipe that specifies how many bcl-2 proteins should be made is altered, and too many bcl-2 proteins are produced. As a consequence of this mutation, B cells that should die, survive. These immortal B cells hang around in a lymph node where they may suffer the additional mutations required to become full-blown lymphoma cells. Because the overexpression of the bcl-2 protein leads to abnormal cell growth (i.e., cells that are immortal), bcl-2 is considered to be a proto-oncogene.

One of the drugs used to treat follicular lymphoma is a monoclonal antibody called Rituxan. This antibody binds to a protein, CD20, which is present on the surface of most follicular lymphoma cells, but which is not found on the surface of blood stem cells or fully mature B cells. When Rituxan binds to CD20, it tags lymphoma cells for destruction by the immune system.

So far, the type of mutation which has been identified most frequently in blood cell cancers is a translocation. However, this is probably an accident of history which arose because a known risk factor for blood cell cancer is exposure to ionizing radiation, and such radiation is known to cause translocations. Further, translocations are relatively easy to spot, because translocated chromosomes frequently look very different from normal ones. Techniques are now available which can identify mutations that are not so obvious, and genetic alterations associated with blood cell cancers that do not involve translocations are being discovered.

If you are a wannabe cancer cell, starting out as a blood cell is probably your easiest route to success. After all, blood cancer cells don't need to learn to metastasize: They can just settle down in the bone marrow or the lymphatic system and work their evil. In addition, since cells in the marrow and the lymphatic system are richly supplied with nutrients, continued proliferation of cancerous blood cells does not require the disruption of systems that control the growth of new blood vessels. Consequently, the number of control systems that must be corrupted to turn a blood cell into a cancer cell is fewer than are required to produce a solid tumor (e.g., a breast cancer). This is probably one reason why leukemia and lymphoma can afflict young people. This is in contrast to most other cancers, which generally develop after the fifth decade of life.

BREAST CANCER

Breast cancer is the most common malignancy of women. Indeed, about one in nine women in America will develop breast cancer, and about half of these will die from the disease. Breast cancer provides an excellent illustration of the important concept that cancer results when growth-promoting and safeguard systems present in normal cells are corrupted. To fully appreciate this point, we need first to talk a little about how breasts develop normally.

Normal Breast Development

Both males and females are born with breasts that can be described as a hunk of fat enclosing a simple plumbing system. The "pipes" that make up this system are the "ducts" that eventually will collect the milk of a lactating mother and channel it to the nipple.

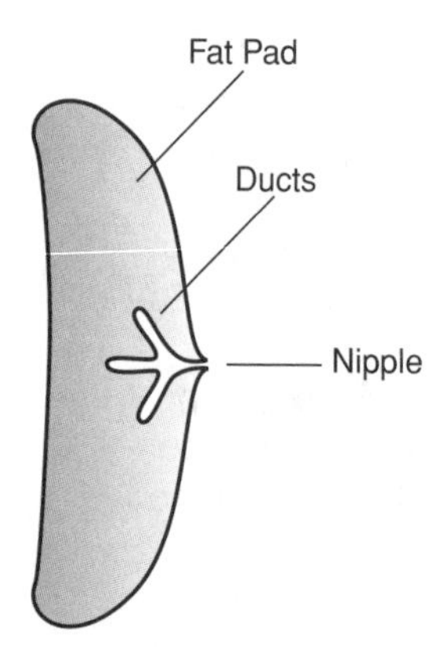

Each duct is made up of a layer of "epithelial" cells that are attached to a membrane called the basement membrane. We need to keep an eye on these epithelial cells, because they are the ones that become cancerous in the majority of breast tumors.

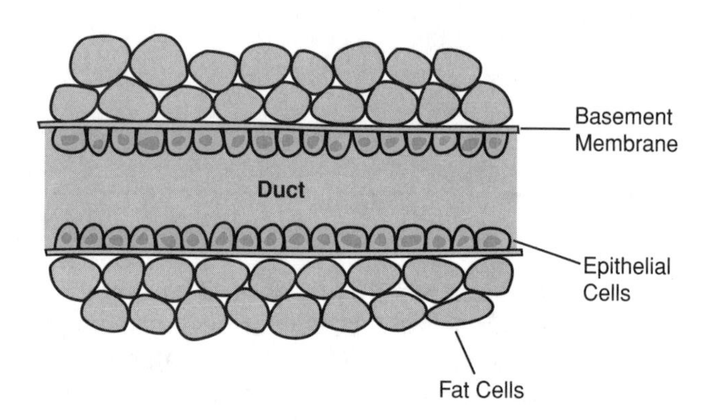

At puberty, under the influence of ovarian hormones (e.g., estrogen), the ends of the ducts that are farthest from the nipple swell to make "buds," and these buds can then split to expand the plumbing system.

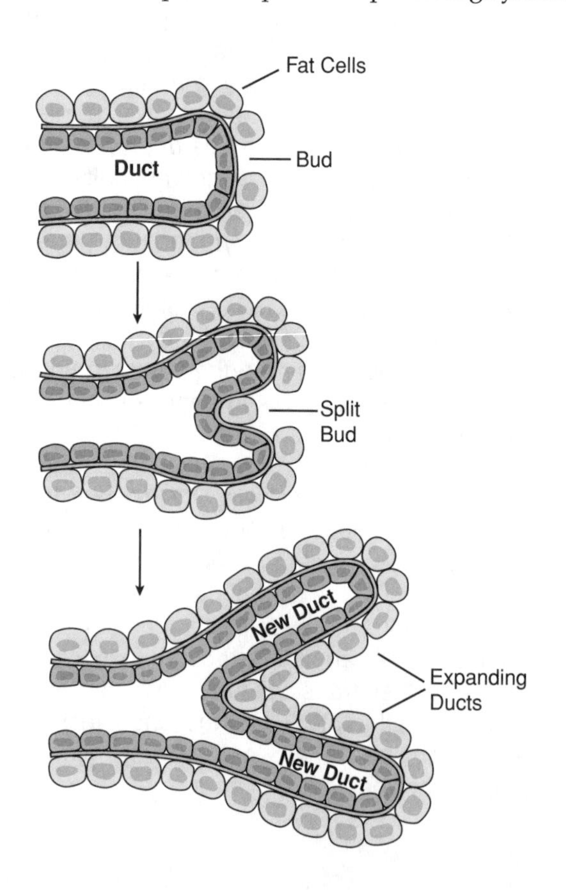

The first thing to note here is that for the ductal system to expand, the epithelial cells that make up the ducts must proliferate. So proliferation of ductal epithelial cells is an essential part of normal breast development. Of course this proliferation must be carefully controlled so that the new ducts don't leak or get clogged up with extra layers of epithelial cells. During the years after puberty, this controlled proliferation continues until essentially the whole breast is "plumbed" with ducts.

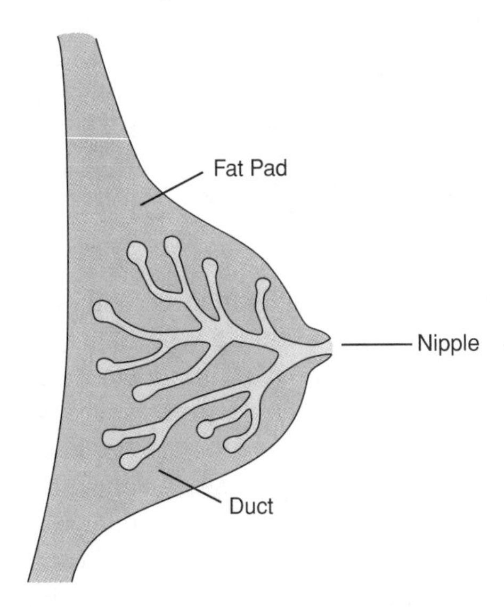

The carefully regulated development of this ductal system is truly amazing—so amazing that, although biologists know that the signals which trigger ductal epithelial cells to proliferate are hormonal, and that "don't proliferate" signals probably are delivered to the epithelial cells when the ducts reach the periphery of the breast, they still haven't figured out how all this is coordinated. Clearly Mother Nature is much smarter than we biologists are!

The next stages of breast development occur during pregnancy. Again, under the influence of estrogen, ductal epithelial cells proliferate, and the ductal system expands even more. This expansion happens in two ways. The first is the extension of the ends of the system by the "branching of buds" that we just discussed. In addition to this branching, the pipes of the system can "sprout" new "side ducts." To accomplish this feat, epithelial cells that line the ducts must break through the basement membrane which supports them and "invade" the fat that surrounds the ducts.

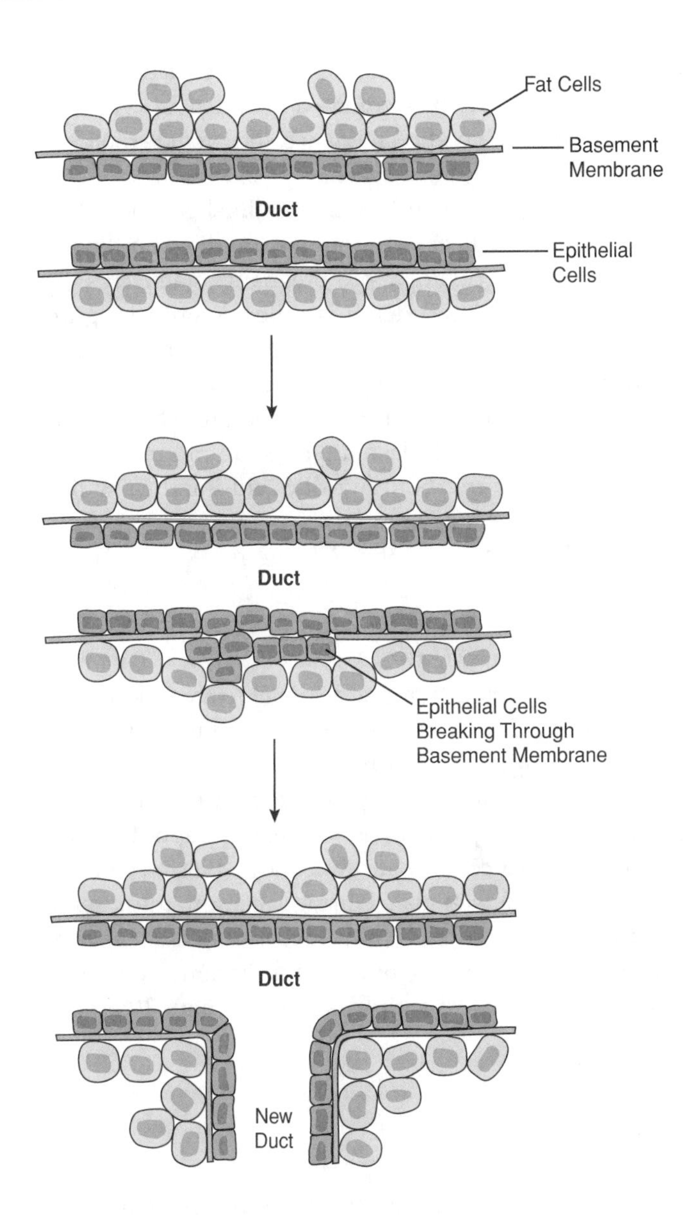

Eventually, as a result of all this branching and sprouting, the breast of a pregnant woman becomes filled with a system of ducts and bulbs. During lactation, under the influence of reproductive hormones, some of the ductal epithelial cells begin to produce milk, and this milk is carried by the ducts to the nipple.

Of course, as the ductal system expands, new blood vessels are needed to support the nutritional needs of cells in the ducts. To encourage the growth of these vessels, substances (e.g., proteins) are given off by the proliferating epithelial cells to help "remodel" the blood supply in the breast so that it can provide nourishment to the cells in the branching and sprouting ducts. Substances that encourage the growth of new vessels are called angiogenic factors, and the process of growing new blood vessels is called angiogenesis. Interestingly, the ability to produce angiogenic factors is not unique to epithelial cells. Indeed, most cells in the body can give off these substances when they are starved for oxygen—a situation in which the growth of new blood vessels that could provide more oxygen would clearly be of benefit.

Finally, when lactation is finished, the cells in most of the ducts are programmed to die, and the ductal system returns to a "pre-pregnant" state. This type of programmed cell death is called apoptosis (gratuitous aside: This is a word some biologist made up because he wanted it to sound Greek!). We saw an example of apoptosis in the last lecture when we discussed the fact that plasma B cells are programmed to die when they are about five days old—and we will see many more examples of this process. Apoptosis is Mother Nature's way of getting rid of cells that are defective or that have passed their prime. Cells which commit suicide in this way are quickly "eaten" by scavenger cells, so they vanish without a trace. This rapid cleanup is certainly a good thing: At lactation's end, programmed cell death is massive, and roughly 90% of the ductal system is "pruned back."

Biologists actually believe that the default option for ductal epithelial cells is death by apoptosis, and that for these cells to survive, they must receive hormonal "rescue signals." During lactation, these rescue signals render the cells of the expanding ductal system resistant to apoptosis, but when lactation ends, rescue signals are in short supply, and only about 10% of the cells survive.

Breast Tumor Development

About 90% of all breast tumors occur when one of the epithelial cells that line the inside of the ducts becomes cancerous, so we'll focus our attention on this type of breast cancer. Although it has been difficult to prove experimentally, most cancer biologists believe that breast cancer cells progress through stages on their way to becoming full-blown tumor cells.

Intraductal Hyperplasia and Carcinoma in situ

Breast cancer begins when one of the ductal epithelial cells loses growth control, and begins to proliferate inappropriately, adding additional layers of cells to the inside of a duct. This condition is called intraductal hyperplasia. Hyper, of course, means "in excess," and plasia comes from a Greek word meaning "to form." So intraductal hyperplasia can be translated as the excess formation of cells on the inside (intra) of a duct.

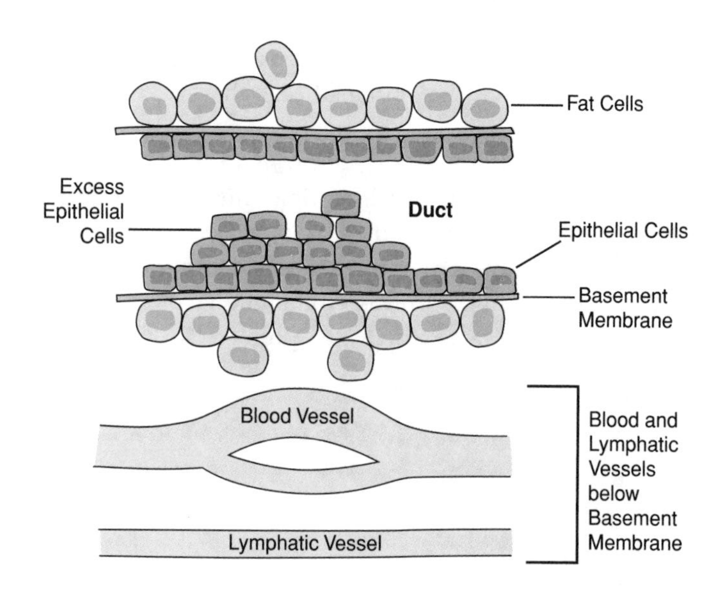

If this uncontrolled proliferation continues, a section of the duct can begin to fill up with deranged cells, leading to a condition known as intraductal carcinoma *in situ*. Carcinoma is the term oncologists use to describe a cancer that arises from one of the epithelial cells that form the sheets of cells (e.g., the skin or the lining of the respiratory tract) which protect our bodies from the outside world. Most human cancers are carcinomas. It may seem odd that breast cancers are called carcinomas. After all, the cells that give rise to these tumors clearly are inside the breast. However, the epithelial cells that line the milk ducts really are exposed to the outside world—through the nipple.

All of the epithelial cells in our bodies, including the epithelial cells that line the ducts in the breast, sit on basement membranes made of collagen fibers. Carcinoma *in situ* (which can be translated as "in place") gets its name because at this early stage, the cancer cells still have not broken through the basement membrane to invade the tissues below.

Infiltrating Ductal Carcinoma

As long as breast cancer remains *in situ*, it is not a problem. Indeed, most ductal carcinomas *in situ* never progress to full-blown breast cancer. The reason is that these cells are located on top of the basement membrane, and the blood vessels they need for nourishment are located below this membrane. As we discussed in Lecture 1, all cells are nourished by molecules (e.g., oxygen and glucose) that are released from blood vessels. No living cell can exist more than about the thickness of a fingernail away from a blood vessel, because at greater distances, the nutrients supplied by the vessels become too dilute, and the cell starves to death. Consequently, because the cells involved in carcinoma *in situ* are separated from the blood supply by the basement membrane, they can only proliferate until they reach a thickness of about ten or twenty cells. At this point, the cells that are on the "top of the stack" begin to starve. Because the growth of these incipient cancers is severely limited by the availability of nutrients, carcinomas *in situ* are so small that they do not even create a detectable lump in the breast.

If additional control systems are compromised in one of the cells that makes up a carcinoma *in situ*, this cell may "figure out" how to break through the basement membrane and escape from the confines of the duct. When this occurs, the cancer is described as infiltrating ductal carcinoma, because at this stage, the cancer cells are free to infiltrate the tissues that lie below the basement membrane. Now things start to get more dangerous.

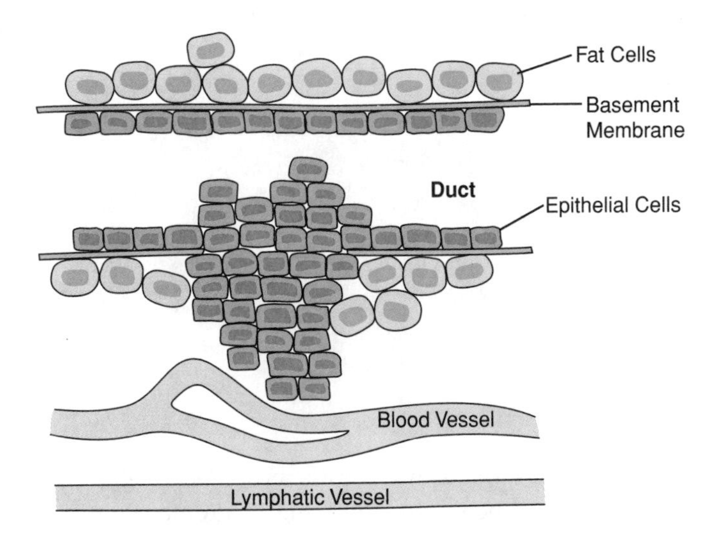

It's important to note here that the growth-promoting systems which must be activated for a cancer cell to proliferate and penetrate the basement membrane are not "abnormal." They are the same systems that normal breast cells use when they penetrate the basement membrane as they sprout new side ducts. Consequently, progression from carcinoma *in situ* to infiltrating ductal carcinoma results from the inappropriate activation of perfectly normal growth-promoting systems.

In the tissues beneath the basement membrane, cancer cells can get right next to blood vessels. More importantly, if additional control systems are corrupted in one of these infiltrating cells, this cell may begin to produce angiogenic factors that recruit new blood vessels into the growing mass of proliferating cells. One of the most famous angiogenic factors, a protein known as VEGF, is given off by breast cancer cells. VEGF is actually a member of a family of closely related proteins—proteins which can stimulate the growth of new blood and lymphatic vessels. Once infiltrating breast cancer cells have access to a blood supply, they can proliferate to form a tumor of considerable size. Because a woman can detect a lump in her breast that is roughly half an inch in diameter, and because an infiltrating ductal carcinoma can easily become this large, it is at this stage that many breast cancers are first detected.

Metastatic Breast Cancer

As long as infiltrating ductal carcinoma remains confined to the tissues of the breast, there is a good chance that the cancer can be cured by removing the tissue that contains the tumor. Where things really get ugly is when, as a result of additional mutations, control systems in one of the invasive carcinoma cells are corrupted so that the cell develops "tools" that allow it to enter the blood or lymphatic system (or both), and spread (metastasize) to other parts of the body. How a cell accomplishes this feat isn't well understood, but some of the same growth-promoting systems that allow the cell to cut its way through the basement membrane probably also help the cell invade the blood and lymphatic systems.

Some cancers seem to favor metastasis via the blood, and others appear to metastasize primarily via the lymph, but all this is still rather mysterious. One aspect that makes this step in metastasis difficult to define is that cells which enter the lymphatic system are collected with the lymph and dumped back into the blood system just before the blood enters the heart to be recirculated. Consequently, whether a cancer cell chooses to invade a blood vessel or a lymphatic vessel, that cell usually will enter the blood circulation rather quickly.

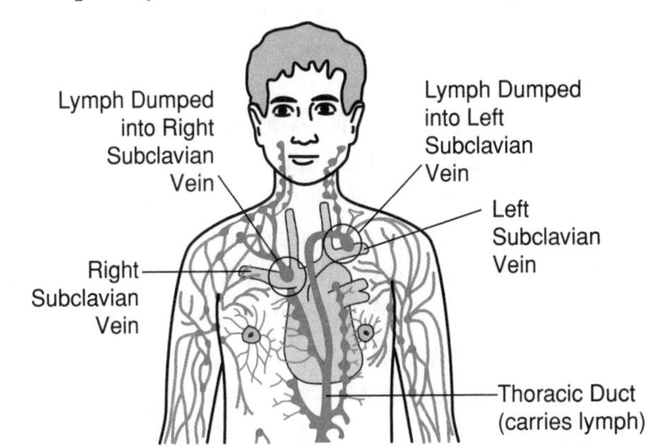

Exactly what happens when a metastatic cancer cell reaches its new home has been very difficult to study. After all, we are talking about what happens to a single cell buried deep within the trillions of other cells that make up our bodies. However, experiments suggest that metastatic cancer cells may experience several different fates, each of which can eventually lead to a large tumor at some distant site in the body. In one scenario, the cell may just sit there, sometimes for years, before it begins to proliferate again. How it can do this, and what awakens the cell from its slumber, are not known. Also, the conditions for growth may not be optimal for the cell in its new home, and although the cell may be able to proliferate slowly, for a while the increase in the number of cells that results from this proliferation may nearly be balanced by cell death caused by the hostile environment.

Because metastatic cells can lie dormant for long periods, and because these cells may adapt slowly to their new environments, metastatic tumors can take years, or in some cases even decades to evolve. This is a big problem, because it is impossible to know whether or not "evolving metastases" are out there—even after successful surgery to remove a primary tumor. Fortunately, there are many steps required for a cancer cell to metastasize successfully, so metastasis is a relatively inefficient process. On the other hand, a primary tumor usually is composed of millions of cells, so this inefficiency sometimes can be overcome by sheer numbers.

Preferred Sites of Metastasis

Because blood flows to every part of the body, it would seem that once cancer cells get into the blood stream, they would go everywhere. However, this is usually not the case: Different cancers (e.g., breast vs. lung) generally have certain areas within the body to which they prefer to metastasize. For breast cancer, the favorite destinations are bones, lungs, and the liver. There are probably two main factors that influence "site selection" during metastasis. The first factor is ease of transport.

When blood containing breast cancer cells enters the heart, it is first pumped through the lungs where it picks up oxygen. So a lung is the very first "stop" on the breast cancer cell's route. Within a lung, the artery bringing in the blood branches into smaller arteries (called arterioles), and these branch even further into smaller vessels called capillaries. These capillaries are so thin-walled that oxygen can freely enter the capillaries from the lungs.

Once blood has been oxygenated in the lungs, it goes back to the heart, and is pumped around the body via arteries. Throughout the body, arteries carrying blood branch into arterioles which then branch further into capillaries. In these capillaries, red blood cells release the oxygen they captured in the lungs, and oxygen and other small molecules (e.g., nutrients) pass out of the capillaries into the tissues. The blood in the capillaries is then collected into veins, and is returned to the heart.

What's important for our discussion here is that most capillaries, either in the lungs or in other areas of the body, have very small diameters—in the range of .003 to .008 millimeters. Red blood cells are able to wriggle through these capillaries because they have a diameter of about .007 millimeters, and because they are built to be very squishy. In contrast, most cancer cells have a diameter of about .02 millimeters—way too large to pass easily through capillaries. Moreover, in contrast to red blood cells, cancer cells tend not to be very flexible, and therefore have a hard time doing the wriggle maneuver. The consequence is that most cancer cells get stuck in the first capillaries they enter.

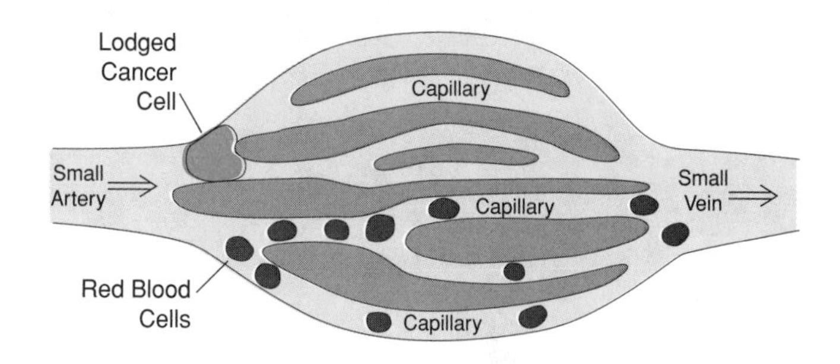

Once lodged in a capillary, a cancer cell is in position to escape the blood system and enter the surrounding tissues.

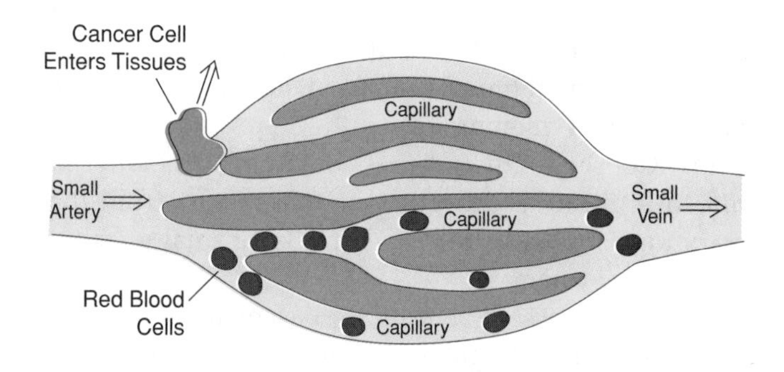

So to get to their "adoptive homes," metastatic cancer cells must break through the basement membrane that encloses their "primary residence," invade the blood or lymphatic circulation, be carried to distant capillaries, extricate themselves from the confines of these capillaries, and enter the surrounding tissues.

The first capillaries which breast cancer cells encounter are in the lungs, so ease of transport is certainly one reason why breast cancer frequently metastasizes to the lungs. Bones are also richly supplied with capillaries, and breast cancer cells that make it past the lungs (probably about 20% of them) have the possibility of "getting stuck" in the capillaries inside our many bones. Finally, about 25% of the blood that leaves the heart with each heartbeat goes through the liver, so the capillaries that nourish the liver are also easily reached by breast cancer cells.

Although ease of transport is certainly one reason why breast cancer cells tend to metastasize to the lungs, bones, and liver, this can't be the whole story. After all, once breast cancer cells enter the blood stream, there are many other organs through which they can circulate. Indeed, it is believed that the environment encountered in its "new residence" probably has as much influence over the choice of metastatic site as does ease of transport—although what "environment" means in this context is far from clear. Every cell in your body depends on the other cells in its "neighborhood" to grow and survive, and the neighborhood of a ductal epithelial cell in the breast is clearly very different from that which a metastatic breast cancer cell will encounter in the tissues of a lung, or a bone, or the liver. Consequently, it is not enough just to plant the "seed" in some distant organ. The "soil" in that organ must also be compatible with the continued growth of the cancer cell. So the two main factors thought to be responsible for site selection during metastasis are ease of transport of a metastatic cell to its adopted home and the compatibility of the environment in this new location with the particular needs of the metastasizing cell.

Metastasis to Bones

When ductal carcinoma cells metastasize to bones, they can interfere with the normal process of "bone remodeling." Each year, about 10% of our bone structure is destroyed and replaced, so about every decade we get a whole new set of bones. To make this happen, two types of cells are at work: cells that destroy old bone and cells that produce new bone. Obviously this remodeling has to be carefully controlled. Otherwise we'd end up either "boneless" or "all bone." When cancer cells metastasize to bones, they upset the balance of the remodeling process, usually leading to bone loss and fragile bones that fracture easily. Bone metastases also can be very painful, although the reasons for this are not well understood.

The Result of Metastasis

When breast cancer cells metastasize to the lungs (about 70% of the cases of metastatic breast cancer) or to the liver (also about 70%), the tumors can become so large that they interfere with the normal function of these organs, resulting in death. So patients with breast cancer don't die because of the primary tumor in the breast. They die when cells from this tumor metastasize to distant sites in the body. Because multiple control systems must malfunction to produce a metastatic breast cancer, it could be predicted that a considerable period of time would usually be required to accumulate all the mutations needed to create these disruptions. Indeed, in contrast to leukemia and lymphoma, which frequently strike during the first five decades of life, about 80% of all breast tumors are diagnosed after the age of fifty.

Metastasis vs. Normal Breast Development

If you compare normal breast development with the evolution of metastatic breast cancer, you will realize that many of the processes required for the development of a normal breast also are required for breast tumor formation and metastasis: proliferation, angiogenesis, resistance to apoptosis, and tissue invasion. This underscores the important principal that the processes that lead to a cell becoming cancerous are perfectly normal cellular processes. Cancer cells don't have to invent anything new. After all, what could be more normal than pregnancy and lactation? The difference between breast tumor formation and normal breast development is that during the evolution of breast cancer, normal cellular processes occur at inappropriate times and places, due to the corruption of cellular systems that control proliferation, angiogenesis, resistance to apoptosis, and invasion.

Cancer Cell Development is Clonal

You may have noticed that as I described how breast cancer progresses from carcinoma *in situ* to metastatic cancer, I said that the corruption of control systems in "a cell" was responsible for the cancer advancing from one stage to the next. Indeed, the picture you should have is that in carcinoma *in situ*, a single epithelial cell loses growth control and proliferates to form a clone of cells, each of which inherits a "growth advantage": the ability to continue to proliferate under conditions in which normal cells would not. Next, one of the cells in this clone of proliferating cells may suffer a mutation that activates a system that controls the cell's ability to penetrate the basement membrane. This cell now has a growth advantage over its "clone-mates," because it can infiltrate the tissues beneath the basement membrane where nutrients are more plentiful. As it proliferates there, it forms a new clone of cells—cells in which two types of control systems have been corrupted: systems which oversee proliferation and systems which control tissue infiltration.

Next, one cell in this clone may mutate so that a system that regulates the production of angiogenic factors is activated inappropriately. This mutation confers a growth advantage on this cell and its progeny, because by recruiting new blood vessels into the growing tumor mass, these cells (which now have three types of corrupted control systems) can gain access to the nutrients required for continued proliferation. Finally, one of the cells in the growing tumor may suffer mutations that allow it to flee the crowded conditions within the primary tumor, and travel to other parts of the body where the "grass is greener."

The important concept here is that each step in the development of a metastatic cancer occurs when, within one cell, corruption of growth-control systems allows that cell to out-proliferate its neighbors to form a new clone. So cancer arises from successive clones of cells, each of which is composed of cells that have a growth advantage relative to cells in the preceding clone.

Genetic Predisposition to Breast Cancer

It is very clear that the genes we inherit can play a role in determining whether we will get breast cancer. For example, if you are a woman whose mother or sister was diagnosed with breast cancer before she was fifty, your risk of contracting breast cancer is increased three- to four-fold. Although it is clear that our individual genetic makeups can help determine whether we will get most types of cancer, breast cancer is one of the malignancies for which important "susceptibility" genes have now been discovered.

To identify genes which, when mutated, might predispose a person to cancer, scientists examine the chromosomes of members of families that are afflicted with an unusually large number of a particular kind of cancer. What they look for are mutated genes that members of these "high-risk" families have in common. This type of search requires the use of very sophisticated techniques and lots of patience, because each of us has about 35,000 different genes. It's like looking for a needle in a haystack.

In the early 1990s, researchers examined DNA from families in which women contracted breast cancer at an unusually early age. Their reasoning was that since mutations in about five control systems are required to turn a normal cell into a cancer cell, a woman who inherits a mutated gene which increases her cancer risk might start life with "one down and four to go." Consequently, these women would be expected, on average, to get breast cancer earlier in their lives than would women who had to wait around for all five mutations to occur. This strategy paid off, and two different "needles" were found in the "haystack" of human genes—genes that were appropriately named BRCA1 and BRCA2.

Only about 3% of the population has mutations in either BRCA1 or BRCA2, but for those who do, there is about a 60% chance that they will be diagnosed with breast cancer during their lifetimes. This is about five times the likelihood that women with normal BRCA1 and BRCA2 genes will get breast cancer. It is important to note, however, that these two mutations don't "cause" breast cancer, since roughly 40% of the women who inherit mutated BRCA genes will never get the disease. Inheriting these mutations simply increases the probability that a woman will get breast cancer: They make her more "susceptible" to the disease. Since the discovery of BRCA1 and BRCA2, biologists have been hard at work trying to understand what these susceptibility genes do. Although the story isn't complete yet, it is now pretty clear why mutant BRCA1 and BRCA2 genes predispose a person to breast cancer.

The BRCA1 and BRCA2 genes contain the recipes for making two different proteins. Each of these proteins is a component of a safeguard system that oversees the repair of double-strand breaks in chromosomal DNA. This type of DNA damage is particularly dangerous, because if the broken pieces of DNA don't get pasted back together correctly, genes can be lost or badly mutated. What scientists discovered is that when either BRCA1 or BRCA2 is mutated, a system that repairs broken chromosomes is disabled. Without this important safeguard system, additional mutations rapidly accumulate, and this increases the probability that other genes involved in growth-promoting and safeguard systems will be mutated—leading to cancer. So normal BRCA genes "suppress" cancer by safeguarding against potential cancer-causing mutations.

Tumor Supressor Genes

The BRCA proteins are each an excellent example of what biologists call a tumor suppressor protein. A tumor suppressor protein is a protein which, if it ceases to function, contributes to a normal cell becoming a cancer cell. And a gene that specifies a tumor suppressor protein is called (that's right!) a tumor suppressor gene. In the case of the BRCA protein, its function is to help keep the mutation rate under control. If the BRCA tumor suppressor gene is mutated so that the BRCA protein no longer works properly, it increases the chances that the breast cell in which this mutation occurred will become cancerous.

You may be wondering why people who are born with a mutated BRCA gene don't just "self destruct." After all, if every cell in their bodies were susceptible to a high rate of mutation, it would seem that in a relatively short time there would be so much genetic damage that most of their cells would simply cease to function. Fortunately, this doesn't turn out to be the case, and the reason is quite interesting.

The BRCA1 gene is located on chromosome seventeen and the BRCA2 gene is found on chromosome thirteen. We inherit two copies of each of these chromosomes—one from Dad and one from Mom—so every one of our cells has two copies of each of the BRCA genes.

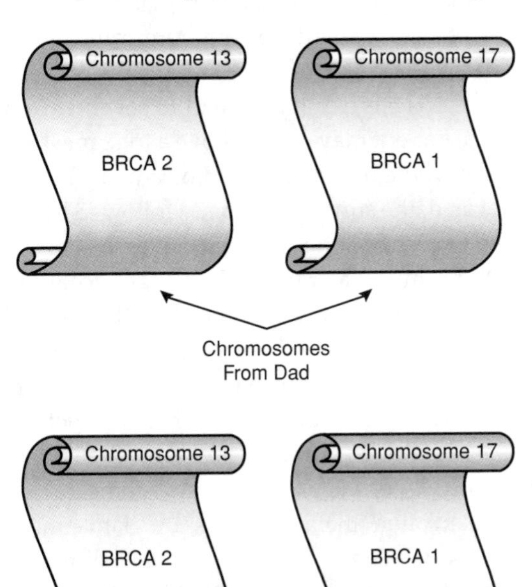

It turns out that if you have one good copy of the BRCA1 and BRCA2 genes, it is enough to make the safeguard system function properly. Only when both copies of a BRCA gene are mutated will the system that repairs broken chromosomes be disabled. This is quite typical. Usually, both copies of a tumor suppressor gene must be mutated before a growth-promoting or safeguard system is compromised.

So the scenario biologists envision is that a woman inherits one bad copy of either BRCA1 or BRCA2. Then, sometime later in life, a mutation occurs in one of her breast cells that corrupts the remaining good copy of that tumor suppressor gene. Once this second copy is mutated, the rate of accumulation of improperly repaired, double-strand breaks increases dramatically in the afflicted cell—because the mutation-suppressing function of the BRCA protein is lost. This can lead to further mutations, which turn on growth-promoting systems inappropriately, and which disable other safeguard systems. If enough of these control systems are corrupted, the result can be metastatic breast cancer.

This hypothesis helps explain why about 40% of women who inherit a BRCA mutation never get breast cancer: They are the lucky ones whose breast epithelial cells either do not suffer a disabling mutation in the second (good) copy of the BRCA gene, or who suffer this "second hit" so late in life that there is insufficient time remaining to accumulate the additional mutations required to produce a breast cancer. This scenario also is consistent with the observation that although most of the cells in the body of a woman who inherits a BRCA mutation contain only the single inherited mutation, both copies of the gene are always found to be mutated in her breast cancer cells.

When a woman inherits a BRCA mutation, that mutation will be present in every cell in her body. However, this mutated tumor suppressor gene does not predispose that woman to leukemia or lung cancer or many other cancers. This brings up an important point: Although multiple control systems must be corrupted to cause cancer, some of the control systems in different cell types are different. Consequently, the genes which must be mutated to cause cancer will differ from cell type to cell type. Certainly, breast cells are very different from blood cells or lung cells. So it makes sense that the growth-promoting and safeguard systems in these different cell types might differ.

Screening for Breast Cancer

Most breast cancers progress very slowly: It usually takes about a decade for a wannabe breast cancer cell to accumulate the mutations required to become a large, metastatic tumor. Consequently, it should be possible to detect breast cancer in the early stages when it can be treated most successfully. Although breast cancers that are about half an inch in diameter or larger can frequently be detected by a physician during a physical examination or by the woman herself, it is not clear that detection at this stage is early enough to make a difference as to whether a woman will die of breast cancer. On the other hand, mammography, which can detect tumors that are about half this size, is a screening method that can reduce the chance of dying from breast cancer, especially in postmenopausal women.

Mammography is essentially a breast x-ray in which the breast is sandwiched (ouch!) between the x-ray machine and the film that displays the x-ray image.

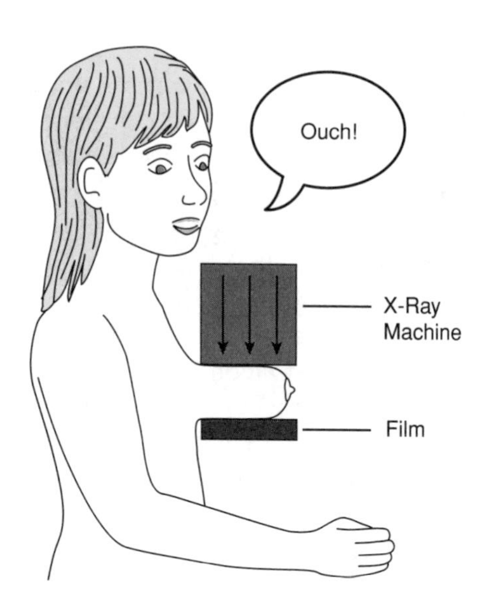

The benefit of mammographic screening for women under fifty is still controversial. However, for women over fifty who would eventually be diagnosed with breast cancer, most studies indicate that early detection by mammography can reduce deaths by about 30%. So for women in this age group, yearly mammograms can be quite useful. Mammograms detect differences in density, and the density of a "lump" is very similar to the density of the tissues found in the breasts of young women. In contrast, the breasts of older women contain more fat, and the density of fat and a cancerous lump are quite different. Consequently, it is easier to detect cancers by mammography in the breasts of older women. In addition, it is thought that the types of breast cancer that younger women get usually progress more rapidly (perhaps because of inherited mutations) than do cancers that are typical of postmenopausal women. So even when a mammogram detects a small cancer in a young woman, it is more likely to have spread outside of the breast.

Women who have a familial history of breast cancer can be tested to determine whether the BRCA genes they inherited are mutant or normal. Based on this information, they can decide whether mammographic screening should be started at an earlier age, and if other measures should be taken to deal with the increased risk conferred by these mutant genes.

Treating Breast Cancer

Treatments for breast cancer are of two basic types: local and systemic. Local treatments are designed to remove the primary tumor from the breast. Systemic therapies are used to treat breast cancer which may have metastasized outside the breast.

Surgery and Radiation Therapy

The most common local treatments for breast cancer are surgery and radiation therapy. For small tumors, a limited amount of breast tissue can be surgically removed—a procedure known as a lumpectomy. On the other extreme, if the cancer is large or has spread to several areas of the breast, the patient may elect to have a total mastectomy in which the whole breast is removed. When a lumpectomy is performed, the surgery is usually followed by radiation therapy. This radiation is focused on the area just around where the tumor was removed, so it is a localized treatment intended to kill any cancer cells in the breast that the surgeon may have missed. This type of "cleanup" seems to be important: The twelve-year recurrence rate for lumpectomy alone is about 35%, whereas when localized radiation therapy is administered after the lumpectomy, only about 10% of the breast cancers "come back."

Radiation therapy works by damaging cellular DNA. This damage is frequently in the form of double-strand breaks, resulting in a cell that is full of broken chromosomes. If a cell that has been irradiated is not proliferating (as most breast cells are not), these double-strand breaks may be reparable, allowing the cell to survive radiation therapy. In contrast, rapidly proliferating cancer cells may divide to produce daughter cells before radiation-induced damage can be repaired. When cell division occurs without repair of broken chromosomes, daughter cells may end up with too few or too many chromosomes. And having the wrong number of chromosomes can be deadly to a cell.

The increased sensitivity of proliferating cells to radiation damage is vividly illustrated by the fact that the brain of an adult, in which very few cells are proliferating, is the organ that is most resistant to damage by radiation. In contrast, the brain of a fetus is the organ that is most sensitive to radiation, because at that stage in development, the fetal brain contains mostly rapidly proliferating cells. So cancer cells that are proliferating rapidly are more sensitive to radiation therapy than are normal cells which are not proliferating. In addition, control systems which can sense DNA damage and inhibit the proliferation of normal cells frequently are compromised in cancer cells. As a result, whereas damaged normal cells usually stop growing while they try to repair damaged DNA, radiation-damaged cancer cells tend to continue proliferating, often resulting in cell death.

Chemotherapy

Local therapies are quite effective in removing or destroying primary tumors. However, it is becoming increasingly clear that by the time breast cancers are detected, many already will have metastasized. To try to deal with the possibility that her tumor has metastasized, a woman may choose to undergo some form of systemic therapy—therapy that is intended to destroy or at least to slow the growth of metastases. Chemotherapy is the most common form of systemic therapy, and for breast cancer, it is conceptually similar to the chemotherapy we discussed for the treatment of leukemia and lymphoma—although different drugs may be used. However, chemotherapy can be quite debilitating, and it is also difficult to tell just which patients will be helped by chemotherapy. Surgeons frequently remove one or more of the lymph nodes that are closest to the breast to try to determine whether metastasizing cells have made their way into the lymphatic system. But even this type of information is not very useful in deciding whether chemotherapy will be effective.

Estrogen and Breast Cancer

In contrast to standard chemotherapy, which is used to treat many different types of cancer, several systemic therapies have now been devised which are tailor-made for breast cancer. One of the most effective of these involves the administration of a drug called tamoxifen. To understand how this drug works, we need to talk a bit about the connection between estrogen and breast cancer.

During normal breast development, the proliferation of ductal epithelial cells is triggered by the hormone, estrogen. Estrogen is a small chemical molecule that can pass right through the membrane that protects each of our cells from the outside environment. Once inside a cell, estrogen binds to a protein called the estrogen receptor, and this binding changes the shape of the receptor. After assuming its new shape, the receptor protein can interact with the control regions of genes that are part of the growth-promoting systems of breast epithelial cells.

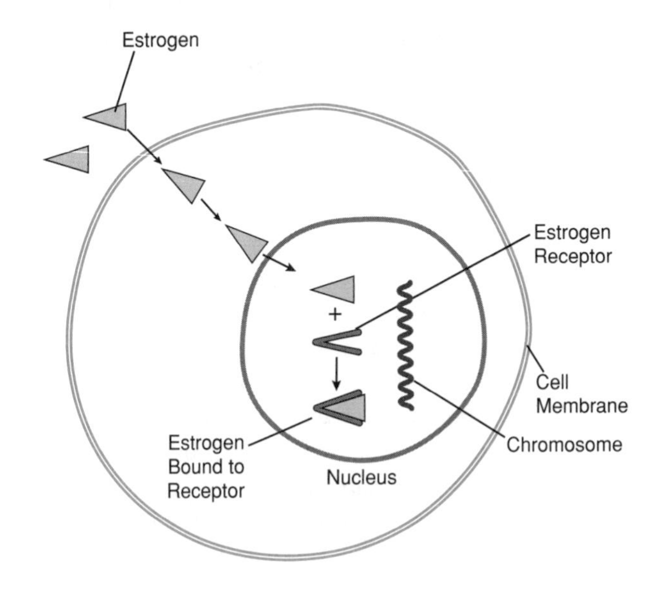

This interaction alters the genetic recipe so that the proteins specified by these "estrogen-responsive" genes are produced in greater abundance, triggering the breast epithelial cell to proliferate. Of course, not all cells in the body make estrogen receptor proteins. In fact, very few cell types do. That's why your big toe doesn't grow in response to estrogen—the cells in your toe don't have estrogen receptors.

Given estrogen's ability to activate growth-promoting systems in normal ductal epithelial cells, and the fact that estrogen is mainly produced by the ovaries, it is not surprising that over 100 years ago it was observed that removal of the ovaries was useful in treating some younger women who had metastatic breast cancer. However, most breast cancer occurs in women who are post-menopausal, and the ovaries of post-menopausal women produce very little estrogen. Consequently, for these women, removing their ovaries would not be very helpful. In addition, although the major source of estrogen is the ovaries, other tissues in the body can also produce this hormone. In fact, some breast cancer cells actually produce their own estrogen! Because there are multiple sources of estrogen, scientists have put a lot of effort into identifying drugs that block the action of this hormone. Tamoxifen is just such a drug.

Tamoxifen

Like estrogen, tamoxifen enters cells easily and binds to estrogen receptor proteins. However, tamoxifen does not change the shape of the estrogen receptor in a way which allows it to turn on estrogen-responsive genes. So in effect, tamoxifen "soaks up" the estrogen receptors, effectively blocking the action of estrogen, and stopping estrogen-induced cell proliferation. But there's a catch: Only about two thirds of women with breast cancer have tumors that produce estrogen receptors, and only about half of these respond to tamoxifen treatment. This illustrates the important concept that all cancer cells are not created equal: Breast cancer is actually a "family" of cancers with different underlying mutations and different properties. Indeed, cancer should be viewed as a collection of closely related diseases that arise in different organs as the result of the malfunction of different components that make up growth-control and safeguard systems.

For breast cancers that are responsive to tamoxifen, treatment with this drug can increase the rate of survival for women who have had breast cancer surgery, can lower the incidence of breast cancer in women from high-risk families, and can decrease the chance that a woman who has had cancer in one breast will get cancer in the other. Indeed, it is estimated that more than 400,000 women are alive today because of tamoxifen. The story of how tamoxifen was developed as an anti-cancer drug is quite interesting.

Back in the 1960s, scientists were searching for a drug that could be used as a "morning after" birth control pill. At that time, birth control was a well-established drug market, and it was likely that such a contraceptive would be a big seller. One of the drugs chemists synthesized in the lab seemed, at least in animals, to have some of the right properties for a contraceptive, but it also had some toxic side effects. So the chemists went to work to try to change its structure slightly to make it less toxic—and the drug they came up with was tamoxifen. Oddly, when doctors tested tamoxifen on humans, they found that it induced ovulation, and in fact, it was first marketed in 1973 as a drug to increase fertility!

Fortunately, when the patent application for tamoxifen was filed, in addition to claiming that it might be used to "manage the sexual cycle," some farsighted person added the words "may be useful for the control of hormone-dependent tumours." Indeed, during the early 1970s, experiments done on animals with breast cancer suggested that tamoxifen might be useful as an anti-cancer drug. After further testing, tamoxifen was approved by the Food and Drug Administration in 1977—the first drug ever approved by the FDA for the prevention of cancer. This year, sales of tamoxifen are expected to gross about $500 million.

Taxol

If metastases from a breast tumor become large enough to be detected in other parts of the body, patients can elect therapies that are focused on delaying the growth of these metastases, thereby improving their quality of life. Standard chemotherapy is generally useful in this situation, and it is sometimes combined with agents which are more breast cancer-specific. One of these is a plant extract called Taxol (a.k.a. paclitaxel) which originally was obtained from the bark of the yew tree, but which now can be made in the laboratory. Taxol interferes with one of the mechanisms required for a cell to divide to make two daughter cells, causing cells that are trying to proliferate to die. So like most standard chemotherapeutic agents, Taxol preferentially targets proliferating cells for destruction. Taxol has been useful in treating both metastatic breast cancer and ovarian cancer.

Herceptin

Another, even more breast cancer-specific therapeutic is a monoclonal antibody that recognizes a growth factor receptor protein called Her-2/neu. These receptors are found in greater than normal numbers on the surface of the cancer cells of about 25% of patients with metastatic breast cancer. Scientists reasoned that an antibody which could bind to the Her-2/neu protein might "cover" this receptor, and prevent it from receiving the "grow" signal that was triggering breast cancer cells to proliferate inappropriately. Their hunch was correct, and the antibody they produced, called Herceptin (a.k.a. Trastuzumab), is frequently used together with standard chemotherapy to slow the growth of metastatic breast cancer in women whose cancer cells overexpress the Her-2/neu receptor.

PROSTATE CANCER

Prostate cancer is the second leading cause of cancer deaths in men in this country. This year, roughly 200,000 Americans will be diagnosed with prostate cancer, and about 36,000 will die from the disease. You may be wondering why I have included breast cancer and prostate cancer in the same lecture, since on the surface, it would seem that these malignancies would be about as different as two cancers could be. After all, women don't even have prostates! However, as you will soon see, breast and prostate cancers share a number of common features. To understand these parallels, we need first to look at the structure and function of a normal prostate.

The Normal Prostate

The prostate's job is to produce prostatic fluid. This watery liquid increases the volume of the seminal fluid that contains the sperm, and also makes this fluid more alkaline, helping the sperm function more efficiently. Like the breast, which also produces fluid (milk), the prostate is filled with ducts. These ducts are lined with epithelial cells, some of which—the "glandular" epithelial cells—produce the prostatic fluid which is channeled by the ducts into the urethra as it passes through the prostate.

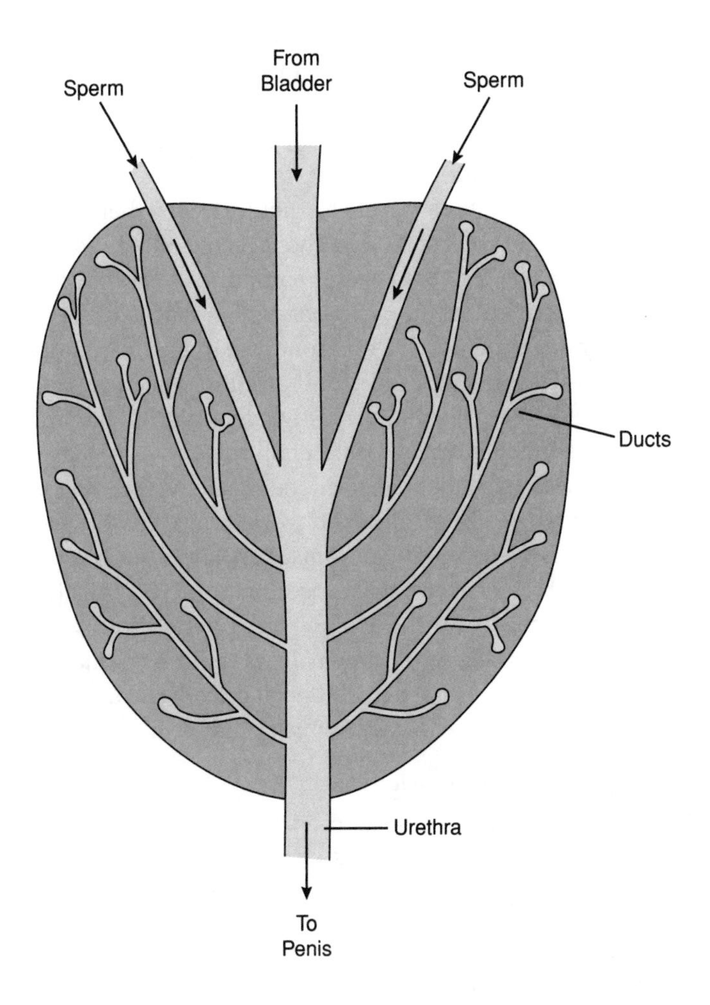

In contrast to the breast, which can be envisioned as a ductal system embedded in fat, the prostate is a ductal system embedded in muscle. This makes sense. Whereas milk is sucked from a lactating breast, prostatic fluid must be "shot" from the prostate—and that's where the muscles come in. During orgasm, these muscles contract and force the prostatic fluid from the ducts into the urethra to help carry the sperm.

The growth of the prostatic ductwork, like the ductwork in the breast, is under hormonal control, and as with most breast cancer, the cells in the prostate that become cancerous are the ductal epithelial cells. However, in contrast to the breasts, where estrogen controls proliferation, the hormones that regulate growth of the ductal system in the prostate are androgens, the most important of which is testosterone. This hormone is produced in the testicles and travels with the blood to the prostate, where it enters the epithelial cells that line the ducts. Once inside these cells, the hormone is modified by cellular enzymes, enters the cell nucleus, and binds to a receptor protein (the androgen receptor). This binding changes the shape of the androgen receptor and allows it to bind to the control regions of "androgen-responsive" genes. This binding changes the genetic cookbook so that the proteins specified by these genes are produced in greater numbers, causing the ductal epithelial cells to proliferate. Although the ductal system in the prostate doesn't expand and contract like the breast ductal system, epithelial cells that line the ducts in the prostate must be replaced as they "wear out." Consequently, a continuous supply of testosterone is required for ductal cell survival and to make this "replacement proliferation" happen.

All this sounds very familiar, doesn't it? The names have been changed from estrogen to testosterone, and different sets of genes are turned on when these hormones bind to their receptors, but otherwise, the hormonal control of ductal cells in the prostate is very similar to the hormonal control of ductal cell proliferation in the breast.

Breast and prostate cancer are both diseases that strike later in life. Breast cancer usually is diagnosed in women who are over fifty, and most prostate cancers are first detected in men who are in their seventies and eighties. With breast cancer, the genes you inherit can influence the chances that you will get this disease, and the same is true of prostate cancer. For example, if either your father or your brother has prostate cancer, the chance that you will develop this disease is increased about three-fold. And if two of your blood relatives (father or brother) have prostate cancer, the likelihood of your getting this type of cancer increases about five-fold. However, in contrast to breast cancer, for which susceptibility genes like BRCA1 and BRCA2 have now been identified, no genes have been found so far which, if mutated, make one more likely to get prostate cancer.

Screening for Prostate Cancer

Two types of screens are commonly used to try to detect prostate cancer at an early stage. The first is the digital rectal exam in which the doctor probes the prostate with his finger to see if it is enlarged or misshapen. Unfortunately, by the time such an abnormality is recognized, many prostate cancers already have metastasized. This situation is reminiscent of the general lack of success in detecting breast cancer before it has metastasized simply by feeling for unusual lumps in the breast.

The second screen for prostate cancer is the PSA test. This test measures the amount of an enzyme (prostate specific antigen) that circulates in the blood. This enzyme is produced by both normal and cancerous prostate cells, and is part of the prostatic fluid that the glandular epithelial cells of the prostate produce. The function of prostate specific antigen is to increase fertility by digesting proteins that can cause sperm to clump together. The PSA test is based on the realization that the epithelial cells of the prostate are the ones which proliferate inappropriately in prostate cancer. Consequently, an increased number of these cells could be expected to result in more PSA in the blood. Most men with a normal prostate will have a PSA level of 4 ng/ml of blood or less.

Although there are exceptions, the PSA test generally can detect the presence of a tumor about five years before it would be recognized by a digital rectal examination. This "lead time" is very important. Before PSA screening became widespread, over 50% of the cancers detected by digital exams had already metastasized and were incurable. In contrast, when prostate cancers are detected in people who have regular PSA tests, most are still confined to the prostate and can be cured.

Treating Prostate Cancer

There are interesting parallels between the ways breast and prostate cancers are treated. With both cancers, early detection is key. If prostate cancer is identified in its initial stages (e.g., by a PSA test), the prostate can be surgically removed or destroyed by radiation. In such cases, the cure rate is quite high: Over 85% of patients with only slightly elevated PSA levels (less than about 10 ng/ml) will survive for longer than five years after surgery or radiation therapy.

One of the difficulties in treating either breast or prostate cancer arises because many of the early stage cancers detected by a mammogram or a PSA test never will progress to become metastatic cancer. After all, multiple control systems must be corrupted before a wannabe cancer cell can become a metastatic tumor cell—and lots of times this just doesn't happen. So a big problem is to know which cancers one can live with, and which cancers one will die from. Because of the uncertainty in predicting whether or not early stage cancers are "dangerous," some breast and prostate cancer patients currently undergo unnecessary treatments. For this reason, biologists are now searching for genetic "markers" in breast and prostate cancer cells that might allow oncologists to make more informed decisions as to who should be treated and who should not.

Once prostate cancer has spread outside the prostate (i.e., has metastasized), it is almost impossible to cure. Prostate cancer most frequently metastasizes to bones—also a favorite site for breast cancer metastases. However, in contrast to breast cancer, prostate cancer almost never spreads to the lungs or the liver. One treatment for metastatic prostate cancer is anti-testosterone therapy. This involves either castration or treatment with drugs (e.g., leuprolide, goserelin, and flutamide) that block either the production or the action of this hormone. Unfortunately, anti-testosterone therapy does not cure metastatic prostate cancer, just as anti-estrogen therapy is not curative for breast cancer. The reason is that prostate and breast cancers both contain a mixture of cancer cells whose proliferation is hormone-dependent and hormone-independent. Anti-hormone treatments can stop hormone-dependent cells from proliferating, and can even cause these cells to die. However, in cells that are hormone-independent, growth-control systems have been corrupted so that the hormone is no longer required for cell growth. These cells simply don't care whether the hormone is there or not—they just keep proliferating.

So men get prostate cancer, and women (mostly) get breast cancer, but many of the characteristics of these two cancers are strikingly similar. In retrospect, this makes sense: Both cancers arise in epithelial cells that are under hormonal control in "glands" which produce fluids.

THOUGHT QUESTIONS

1. Discuss the similarities and differences between normal breast development and the development of a breast tumor.
2. BRCA1 and BRCA2 are excellent examples of tumor suppressor proteins. Explain what these tumor suppressor proteins do.
3. Discuss the steps that are required for breast cancer to metastasize.
4. Breast cancer and prostate cancer have some features that are amazingly similar. Compare and contrast these two cancers.
5. Cancer cell development is clonal. Explain what this means and give examples.
6. Tamoxifen is commonly used to treat breast cancer. Discuss how this drug works.

Table of Concepts for Lecture 4

Concept	Example
Inherited genes can predispose to cancer.	BRCA genes in breast cancer
Cancer results when normal control systems are corrupted. Cells don't need to learn anything new.	Breast cancer
The same cancer can result from different mutations in different control system genes. Each cancer is a "family" of diseases.	Breast cancer
Hormones can trigger cancer cell growth.	Breast and prostate cancer
Tumor suppressor proteins can help repair mutations.	BRCA proteins in breast cancer
Screening procedures can detect certain cancers at an early stage.	Mammograms and breast cancer PSA test and prostate cancer
Cancer cell development is clonal.	Breast cancer
Metastasis is a multi-step process.	Breast cancer
Different cancers usually have favorite sites to which they metastasize.	Breast and prostate cancer
Monoclonal antibodies can be used to treat cancer.	Herceptin for breast cancer
Anti-hormone drugs can treat hormone-dependent tumors.	Breast and prostate cancer

Lung and Skin Cancer

REVIEW

In the last lecture, we examined two seemingly unrelated malignancies that turn out to have interesting similarities—breast and prostate cancer. Both cancers usually originate when multiple control systems are corrupted in one of the ductal epithelial cells that are important components of the "plumbing" in these two organs. Proliferation of ductal epithelial cells in both organs is controlled by hormones—estrogen in the breast and androgens (e.g., testosterone) in the prostate—so it is not unexpected that in both cancers, the hormone-regulated, growth-promoting system is often compromised. Both cancers demonstrate clearly that to become a cancer cell, nothing "new" is required. The systems that are corrupted in breast and prostate cancer cells are systems that are indispensable during the normal development of each organ. These include systems involved in cell proliferation, angiogenesis, resistance to death by apoptosis, and tissue invasion.

With both breast and prostate cancer, the genes you inherit can play a part in determining your chances of getting the disease. Two of the genes that affect susceptibility to breast cancer are BRCA1 and BRCA2. The proteins specified by these genes are components of a safeguard system which oversees the repair of double-strand breaks in chromosomes. If one of the BRCA genes is mutated so that this safeguard system is disabled, the mutation rate increases dramatically, making it more likely that the cell will suffer additional mutations in other control systems. The proteins specified by the BRCA genes are excellent examples of tumor suppressors—proteins which, if they <u>cease</u> to function, contribute to transforming a normal cell into a cancer cell. Fortunately, as is true of most (but not all) tumor suppressors, both copies of a BRCA gene must be mutated before suppressor function is lost. This is because a single good copy of a BRCA gene directs the production of enough BRCA tumor suppressor protein to fulfill its role in repairing damaged DNA.

Although inheriting a mutated BRCA gene significantly increases a woman's chances of getting breast cancer, only about 3% of women who have breast cancer inherit a mutated BRCA gene. This illustrates the important point that breast and prostate cancers are each a "family" of cancers in which different combinations of mutations can result in the corruption of different growth-promoting and safeguard systems. Although the end result is the same—a cancer cell—different combinations of mutations can produce cancer cells with very different properties (e.g., sensitivity or insensitivity to hormones).

In order to treat cancer effectively, it is important to detect cancer cells in the early stages of their development—at a time when they are still confined to the organ in which they originate (i.e., before they have metastasized). For both prostate and breast cancer, screening procedures exist which are helpful in spotting tumors early. Regular mammograms can decrease the probability that a woman over about fifty years of age will die from breast cancer. Likewise, a simple blood test that measures the level of prostate specific antigen can lead to a diagnosis of prostate cancer years before symptoms of the disease would signal its presence. Localized breast and prostate cancer frequently are treated by removing the tissues that include the tumor. For breast cancer, this ranges from fairly limited surgery, in which a lump of breast tissue is cut out, to a total mastectomy, in which the complete breast is removed. For early stage prostate cancer, the entire prostate is usually either removed surgically or destroyed by radiation therapy. Although localized breast and prostate cancer frequently can be treated successfully, once these cancers have metastasized, they rarely can be cured.

Most cancers have preferred sites within the body to which they metastasize. For example, breast cancer normally metastasizes to the lungs, bones, and liver, whereas prostate cancer usually spreads to the bones, but rarely to the lungs or liver. How cancers select the sites to which they metastasize isn't totally clear. However, two important factors that influence site selection are how easily cancer cells can be transported from a primary tumor to their new home, and how compatible the environment at their new location is with the needs of the metastatic cells.

Once metastatic cells reach their destinations, they can lie dormant or grow slowly, sometimes taking years before they adapt to their new surroundings and proliferate to form a deadly tumor. Fortunately, treatments are available for both breast and prostate cancer that can slow the growth of metastases, and can prolong the life of the patient. One approach that has been useful for both cancers is anti-hormone treatment in which drugs are administered that interfere either with the production or the action of estrogen (breast cancer) or testosterone (prostate cancer).

LUNG AND SKIN CANCER

In this lecture, we'll examine another pair of cancers that on the surface would seem not to be very similar. However, lung and skin cancer do have a number of features in common, and an analysis of the similarities and differences between these cancers will reveal important features of cancer in general.

Lung Cancer

Lung cancer is second only to breast cancer in women and prostate cancer in men in the number of people it afflicts. However, lung cancer kills more men and women in the United States than any other cancer. These statistics make the important point that lung cancer is extremely difficult to cure: Only about 10% of lung cancer currently is curable.

Like breast and prostate cancer, lung cancer usually occurs when one of the cells that lines a ductal system becomes malignant. Of course, the primary function of the ductal system in the lungs is not to produce fluids, although fluid (mucus and surfactant) is produced by some lung cells. The ducts in the lungs facilitate the exchange of oxygen between the air we breathe and our blood—so they are really "air ducts."

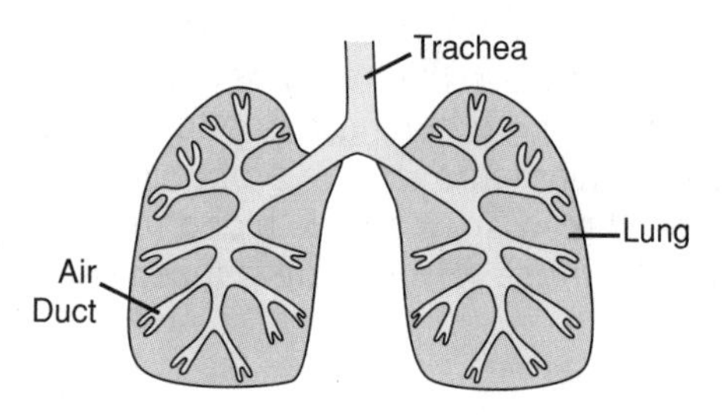

Lung cancers which arise in the epithelial cells that line the air ducts are classified as carcinomas. Lung carcinomas really represent a family of cancers that are grouped, more or less according to the size of the cells which make up the tumor, into two categories: small cell lung cancer and (ah, what shall we call the others?) non-small cell lung cancer. Roughly 80% of lung cancer is of the non-small cell type.

Risk Factors for Lung Cancer

For many cancers, it is known or suspected that "environmental factors" can increase the chances that a person will get the disease. By environmental factors, I mean "outside" influences, as opposed to differences in the genes we inherit. However, for most cancers, the exact nature of the environmental factors that influence cancer susceptibility is difficult to sort out, because there are generally many such factors which individually make only small contributions to the overall cancer risk. For example, the incidence of prostate cancer in China is about fifty-fold lower than in America. Although some of this difference is likely to be genetic, when Chinese move to the United States, their rate of prostate cancer increases to more closely resemble that of the American population. The thinking, of course, is that there is something different between the environment in China and the United States (perhaps the diet or the lifestyle) that affects the risk of getting prostate cancer. Nevertheless, epidemiologists have not been able to discover any single factor that can explain the difference in prostate cancer risks for people living in these two countries. In contrast, it is very clear that three environmental factors, acting either singly or in concert, can dramatically increase a person's risk of getting lung cancer: cigarette smoking, asbestos inhalation, and inhalation of radon gas and its radioactive decay products.

Cigarette Smoking and Lung Cancer

Why is cigarette smoking such an important risk factor for lung cancer? First, the cells in the lungs which become malignant come in direct contact with cigarette smoke. This is in contrast to organs like the breast or the prostate that don't "suck in" anything from the environment. Moreover, cigarette smoke contains more than 4,000 chemicals, and about sixty of these have been demonstrated to be "carcinogens." Exposure to these carcinogens can increase the rate at which cells accumulate mutations in growth-promoting and safeguard systems. Nevertheless, the carcinogens in cigarette smoke do not produce instant cancer: There is usually a lag time of about thirty years between when a person begins to smoke and when he dies of lung cancer. This lag time presumably reflects the fact that the safeguard systems within lung cells are powerful enough to deal with most of the mutations caused by smoking, and that multiple control systems must be corrupted. Only after an extended period will a cell acquire the collection of mutations required to inappropriately activate growth-promoting systems and disable safeguard systems, resulting in lung cancer.

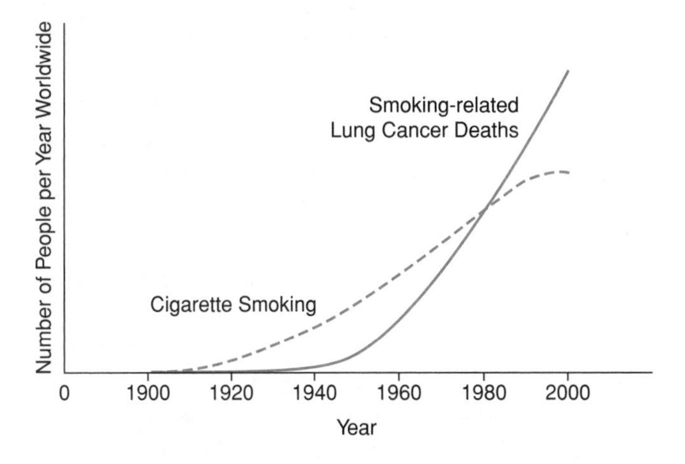

You will notice from this graph that cigarette smoking only began on a large scale at the end of the nineteenth century—after safety matches (1855) and the automatic cigarette rolling machine (1880) were invented. Because of this "late start," lung cancer was an extremely rare disease until well into the second decade of the twentieth century. Indeed, by 1912, fewer than 200 total cases of lung cancer had been reported! Now so many people smoke cigarettes that roughly one million people will die of lung cancer this year worldwide.

About 90% of all people who get lung cancer are cigarette smokers, and it is estimated that the total number of cancer deaths each year from all types of cancer combined could be reduced to about 40% of its present level if nobody smoked. In fact, it could be argued that if we could "make" people quit smoking, we would have a greater impact on cancer deaths than all the money spent on cancer research thus far. But, of course, making people change their lifestyles is notoriously difficult.

Asbestos and Lung Cancer

A second clear risk factor for lung cancer is asbestos inhalation. Asbestos was once billed as a miracle fiber. It is an excellent thermal insulator, and is resistant to flame. Because it could be woven into cloth, it was used to make everything from oven mitts to fireproof suits. Unfortunately, materials made of asbestos shed asbestos fibers, and these are easily inhaled. Asbestos fibers can work their way through the lung and damage the membrane (the pleura) that lines the chest cavity. Although asbestos came into use in the 1880s, only in the late 1950s was it recognized that people exposed to asbestos had an unusually high rate of an otherwise rare cancer of the pleura called mesothelioma. This type of lung cancer is incurable, and usually kills within months of diagnosis. The fact that mesothelioma was rare in the general population should have made it easier to recognize the connection between asbestos and cancer. However, the lag time between exposure to asbestos and the appearance of this cancer can be as long as sixty years, and this made it difficult to connect asbestos inhalation and mesothelioma in a timely fashion. Indeed, because this connection was so slow to be recognized, the incidence of mesothelioma is still increasing in this country.

One of the interesting features of mesothelioma is that although exposure to high levels of asbestos can increase the risk of getting this cancer about seven-fold, for asbestos workers who smoke, the mesothelioma risk is about 50 times that of the general population. So asbestos exposure and cigarette smoking can combine to greatly increase the risk of contracting mesothelioma. This is not uncommon: Frequently, two risk factors can work together (synergize) to create a greater risk for cancer than the sum of the individual risks conferred by these factors. For mesothelioma, the basis for this synergy is not well understood. However, it has been hypothesized that when asbestos in the lungs causes inflammation, some of the molecules released during this inflammatory reaction act as growth factors for cells of the pleura, causing them to proliferate. This forced proliferation may increase the rate of smoking-induced mutation, because proliferating cells have less time to repair the damage to DNA inflicted by the carcinogens in cigarette smoke.

Radon Gas and Lung Cancer

When we think of exposure to ionizing radiation, we typically think of people like Homer Simpson who work at a nuclear reactor or workers at a facility which deals with nuclear waste. However, there is another source of ionizing radiation to which many of us are subjected: radon gas and its radioactive decay products.

Radioactive uranium is found in the soil in every one of our fifty states. It decays first to radium, which then decays to radon. Whereas uranium and radium are solids, which stay put in the soil, radon is a colorless, odorless gas which can leach out of the soil into the environment. Indeed, the national average of the amount of radioactivity emitted by radon gas in the atmosphere is about three Curies per 100 billion liters of air. Every four days, about half of this radon decays into polonium (a solid) which then decays further into other nongaseous isotopes. And each time radon or polonium decays, an alpha particle is emitted. Alpha particles are big trouble. Although they give up their energy quickly and therefore only travel a short distance, they carry so much energy that they can severely damage cellular DNA. If sufficient quantities of radon or its decay products are taken into the lungs, the resulting DNA damage can dramatically increase the risk of lung cancer. Indeed, inhalation of radon and its decay products is second after cigarette smoking as a risk factor for lung cancer. As with asbestos inhalation, exposure to radon gas synergizes with cigarette smoking to further increase the risk of lung cancer. This synergy may result because these two environmental carcinogens produce complementing mutations—or it may simply be that cigarette smoke increases the amount of radon decay products that becomes lodged in the lungs.

Although it is believed that the level of radon in the outside air does not pose a significant health risk, where things get dangerous is when radon gas and its decay products are trapped inside a house. For example, when we measured the level of radon in our home here in Idaho, we found it was about 250 Curies per 100 billion liters of air—roughly 80 times the average outdoor concentration. To deal with this problem, the ground under our house was covered with plastic, and a fan was installed which exhausts to the outside most of the radon that emerges from the soil beneath the plastic. In this way, we were able to reduce the radon concentration in our home below the level of 40 Curies per 100 billion liters suggested by the EPA. It is estimated that roughly 7% of the homes in this country have indoor radon gas levels that exceed this recommended level.

Lung Cancer and Tumor Supressor Genes

In the last lecture, we discussed the concept of tumor suppressor genes and the proteins they specify. Two of the most important tumor suppressors of all, p53 and RB, frequently are mutated in lung cancer cells.

p53

The tumor suppressor gene, p53, was given this name because it contains the recipe for making a protein that has a molecular weight of 53,000. In more than 90% of small cell lung carcinomas, both copies of the gene for p53 have been mutated so that the p53 proteins no longer function. The reason this tumor suppressor is so important is that the p53 protein plays a major role in a safeguard system which senses unrepaired DNA damage, and either stops cell proliferation to allow additional time for repair, or triggers the damaged cell to commit suicide. If the genes for p53 are mutated so that functional p53 proteins are not produced, this "not so fast" safeguard system is disabled, and unrepaired mutations accumulate at a rapid rate.

Every carcinogen produces "signature mutations." For example, ionizing radiation is a carcinogen that frequently breaks both DNA strands. One of the most potent carcinogens in cigarette smoke, benzo(a)pyrene, causes a signature mutation in which one of the "letters" used in the genetic cookbook is altered. The recipes in this cookbook are written in a four-letter code in which each letter (A, G, C, or T) is one of the building blocks used to construct DNA. When even a single letter in one of the recipes is altered—the equivalent of a minor spelling error—the protein that is produced according to this recipe may no longer function. What's interesting about benzo(a)pyrene is that this carcinogen doesn't change just any old letter. It almost always changes a G to a T. Indeed, when mutated p53 genes present in lung tumors of patients who smoke cigarettes are examined, these mutations are mainly G to T changes. In contrast, lung cancer cells from non-smokers have many fewer of these signature G to T mutations. The fact that benzo(a)pyrene leaves a distinctive mark on the genes of cigarette smokers strengthens the connection between smoking and lung cancer.

RB

In over 90% of small cell lung cancers, both copies of a tumor suppressor gene called RB are mutated. This gene was named because it was first identified as being important in a childhood cancer called retinoblastoma. Like p53, RB is mutated in many different types of cancer. The reason for this is that the RB protein is a very important component of a growth-control system that determines whether or not a cell is allowed to proliferate.

When a cell proliferates, it first must copy the DNA in its chromosomes (the genetic cookbook), so that when it divides to make two daughter cells, each daughter will get a complete set of recipes.

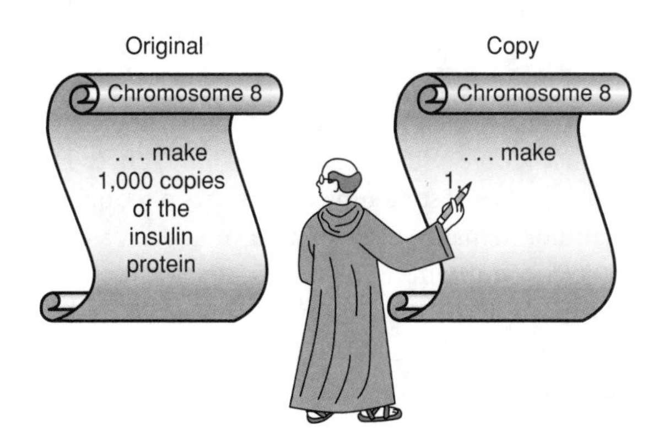

In every cell, there is a protein called E2F which is responsible for turning on a whole battery of genes required for the cell to copy its DNA. In a cell that is not proliferating, the RB protein binds to E2F and keeps it from turning on these genes. As a result, DNA copying can't take place. On the other hand, when growth factors signal that the time is right for a cell to proliferate, an enzyme in this growth factor pathway sticks a phosphate group on the RB protein, changing its shape and causing it to release E2F. Now the E2F protein can do its thing and can turn on the genes needed for DNA copying. In effect, the RB protein (pRB) functions as a "parking brake" which keeps cells from copying their DNA until growth factors signal that proliferation is needed.

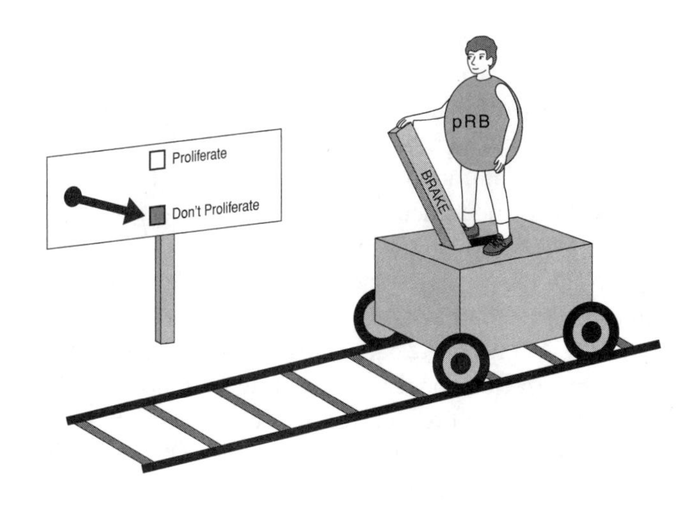

At least, that's the way the pRB parking brake is supposed to work. If, however, the RB gene is mutated so that the protein it specifies no longer can bind to E2F, the E2F protein is then free to initiate DNA copying. Importantly, when RB is mutated in this way, the cell no longer cares whether it is receiving "proliferate" signals from growth factors: The parking brake has been released, and the cell just goes right ahead and copies its DNA. Like most tumor suppressors, both copies of the RB recipe (gene) must be mutated before the brakes are released. This is because the amount of RB protein specified by just one of the two RB genes is sufficient to bind up the E2F proteins in the cell, preventing the inappropriate copying of the cell's DNA.

Interestingly, although RB mutations are found in almost all cases of small cell lung cancer, only about 15% of non–small cell lung cancers have RB mutations. This is another example of the important concept that the mutational path to cancer can be different, even in cancers that arise in the same organ.

Oncogenes vs. Tumor Supressor Genes

We have now discussed two types of cellular genes which, if mutated, can contribute to the development of a cancer cell: proto-oncogenes and tumor suppressor genes. Although oncogenes and mutated tumor suppressor genes both play important roles in cancer, these roles are very different. So we need to take a moment to be sure the difference is clear.

We have encountered several excellent examples of oncogenes. For instance, the oncoprotein specified by the BCR-ABL fusion gene in chronic myeloid leukemia acts to short-circuit a growth-control pathway, tricking the cell into thinking it's receiving "proliferate signals" from outside the cell. A second oncogene, the mutated bcl-2 gene, directs the production of too much bcl-2 protein. These excess bcl-2 proteins interfere with the normal workings of a system that controls when a cell should die, keeps blood cells alive past their normal "expiration" date, and increases the likelihood that these cells will become follicular lymphoma cells.

We have also discussed three tumor suppressor genes. The BRCA genes are important components of a safeguard system that oversees repair of damaged DNA. If the BRCA1 or BRCA2 genes are mutated so that they no longer specify functional proteins, double-strand breaks in DNA are not repaired correctly, the mutation rate increases, and a cancer cell may result. The p53 tumor suppressor protein is part of a safeguard system that detects genetic damage, and stops cells from proliferating until the damage can be repaired. If the genes for p53 are mutated and the function of the p53 proteins is lost, cells are free to proliferate without pausing to repair DNA damage, resulting in mutations that can lead to cancer. Finally, the RB protein acts like a parking brake which keeps cells from copying their DNA until growth factors signal that proliferation is appropriate. If both copies of the RB tumor suppressor genes are mutated so that the RB proteins lose their ability to apply the brakes, the result is uncontrolled proliferation—an important step in becoming a cancer cell.

What proto-oncoproteins and tumor suppressor proteins have in common is that they are components of cellular growth-promoting and safeguard systems. If the genes that specify these proteins are compromised, the systems in which they participate can malfunction, contributing to the development of a cancer cell. Where they differ is that oncoproteins play an active role in compromising control systems, whereas mutant tumor suppressor proteins corrupt control systems when they lose their ability to function.

Until recently, biologists believed that proto-oncogenes and tumor suppressor genes were different in that only one copy of a proto-oncogene needed to be mutated to corrupt a control system, whereas both copies of a tumor suppressor gene had to be inactivated before a control system would be compromised. However, tumor suppressors have now been identified which cease to function when only one of their gene copies "goes bad." So this distinction between proto-oncogenes and tumor suppressors isn't nearly as clear today as it used to be. For me, I like to focus on what the proteins specified by these genes actually do or don't do within the cell—and whether they disrupt a growth-promoting or a safeguard system. That's what really matters. The labels "oncoprotein" or "tumor suppressor protein" really aren't so important.

Treatments for Lung Cancer

One reason lung cancer is so deadly is that, unlike breast or prostate cancer, there is no good screen which can be used to detect lung cancer in its early stages. Indeed, most lung cancers that can be cured are discovered by accident when a patient has a chest x-ray for an unrelated condition (e.g., during a routine physical). Studies have been done to evaluate the effectiveness of using x-ray exams to screen for lung cancer in male smokers, but most results indicate that this type of screening does not reduce the chances that these men will die of lung cancer.

When non–small cell lung cancer is detected early, the standard treatment is to surgically remove the portion of the lung that contains the tumor. Usually, by the time lung cancer produces symptoms (e.g., coughing, wheezing, or chest pain), it will already have metastasized to other organs, making it almost impossible to cure. Non–small cell lung tumors generally metastasize to the small intestine, brain, liver, adrenal glands, and bones. Even for patients who are asymptomatic when their non–small cell lung cancer is diagnosed and whose tumors are removed surgically, only about half will survive as long as five years. After surgery, chemotherapy or radiation therapy appears to be of little value in increasing survival time for most people with non-small cell lung cancer. However, these treatments sometimes are useful in reducing the size of a tumor to make surgery easier or in decreasing the severity of the symptoms of metastatic disease.

Although the prognosis for non–small cell lung cancer patients usually is measured as "expected time of survival," the outlook for patients with small cell lung tumors is even more bleak. For example, it is not clear whether surgical removal of these cancers is of any benefit, because at time of detection, these tumors usually have metastasized to the liver, bones, brain, adrenal glands, lymph nodes, and pancreas.

Although the prognosis for most patients with lung cancer is not good, about 90% of these cancers can be prevented. Just don't smoke!

Weight Loss and Cancer

One of the conditions that can lead to death in cancer patients is a loss of body fat and muscle mass called cachexia. This "wasting" syndrome is especially common in individuals with lung cancer and cancer of the gastrointestinal tract (e.g., colon cancer). Weight loss in cancer patients was originally believed to be due simply to loss of appetite (anorexia), which is a common side effect of cancer-associated pain or cancer therapies.

However, it is now known that weight loss due to decreased food intake and weight loss that results from cachexia are quite different.

If a person suffers from anorexia and takes in fewer calories than are burned, the body begins to utilize stores of fat to obtain the energy it needs. If this continues for a while, the result is a person who has little body fat, but who may still have excellent muscle tone (e.g., a supermodel). It is only when all fat reserves are exhausted that the cells in the body begin to cannibalize their own proteins as an energy source, resulting in the loss of muscle mass.

With cachexia, the sequence of events is quite different. Here a person loses both fat and muscle from the very beginning, and the breakdown of muscle proteins can be life threatening. For example, the muscles that operate the diaphragm may become so weak that the individual no longer can breathe. Indeed, cachexia is responsible for about 20% of all cancer deaths. Research on the biological basis of cachexia is still in the early stages, and it is not yet clear why some cancer patients metabolize both fat and protein. It has been proposed that tumors give off factors that either directly or indirectly influence protein metabolism. Although at least one such factor, a protein called PIF, is found at higher concentrations in the blood of individuals suffering from cachexia, little is known about how these factors work. Interestingly, it has recently been shown that one of the ingredients in fish oil (EPA) or fish oil itself, in conjunction with a high-protein diet, can reduce levels of PIF and can counteract muscle loss in cancer patients with cachexia.

Skin Cancer

Skin cancer is the most common malignancy in humans, and its incidence is increasing more rapidly than that of any other cancer. Like most breast, prostate, and lung cancers, skin cancers arise in cells that form the interface between our bodies and the outside world. I say "cancers" because skin cancer actually comes in three popular flavors: basal cell carcinoma, squamous cell carcinoma, and melanoma.

Basal cell carcinoma is the most common skin cancer. Fortunately, this type of cancer rarely metastasizes: Basal cell carcinomas usually just continue to grow where they originate—in the skin. Squamous cell carcinoma is less common than basal cell carcinoma, but this type of cancer metastasizes more frequently than does basal cell carcinoma, and therefore is more dangerous. However, both basal cell and squamous cell carcinomas pale in comparison to the deadliest skin cancer of all: malignant melanoma. Because melanoma is so deadly, and because the lifetime risk for an American developing this type of cancer is about 1 in 60, we will focus mainly on this disease. The choice of melanoma as our "model" skin cell cancer also makes sense because more research has been done on this type of skin cancer than on its two "siblings"—so more information is available to help us fathom how melanoma works.

To understand skin cancer, we must first understand the structure of skin. Skin can be divided into two parts: the epidermis, which is exposed to the outside world, and the dermis, which lies below it. These two regions are separated by a basement membrane. Attached to the epidermal side of the basement membrane are "basal cells," and scattered among these cells are basal stem cells that proliferate on demand to replace epithelial cells as they are lost or damaged. Some of the progeny of these stem cells remain attached to the basement membrane and become stem cells themselves, whereas others are pushed upward toward the surface of the skin. Once disengaged from the basement membrane, basal cells mature and stop proliferating.

Driven toward the surface by the continued proliferation of basal stem cells, the maturing epithelial cells of the skin dedicate themselves to the production of keratin proteins. Because these cells are keratin factories, they are called keratinocytes ("cyte" is derived from the Greek word, *kytos,* which means cell). As the keratinocytes approach the skin surface, they flatten and eventually die, exhausted from the all-out effort of producing huge amounts of keratin protein. In death, they become flattened bags of keratin. These dead cells, usually ten to twenty layers deep, function as overlapping

Annual New Cases and Deaths in the United States

Cancer Type	Approximate New Cases/Year	Approximate Deaths/Year
Basal Cell Carcinoma	1,000,000	200
Squamous Cell Carcinoma	250,000	2,000
Melanoma	50,000	8,000

"shingles" which provide protection against the outside environment. When, as the result of wear and tear, these dry shingles flake off the skin surface, they are replaced by a new set of shingles rising from below—the end result of basal cell proliferation. Usually it takes several weeks for the descendants of basal stem cells to reach the skin surface and become household dust.

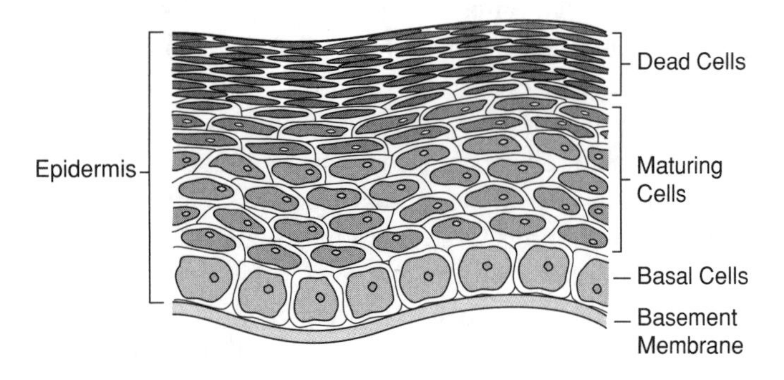

Basal cell and squamous cell carcinomas are derived from basal epithelial cells which normally would mature to become keratinocytes. In contrast, melanomas originate from cells called melanocytes that sit on the basement membrane, shoulder-to-shoulder with basal epithelial cells. Only about 10% of the cells in the basal layer are melanocytes, and melanocytes seldom proliferate. These are amazing cells that specialize in making a pigment protein called melanin, which they package into little sacks that can be transferred to nearby basal epithelial cells. By "gifting" basal epithelial cells with the melanin-containing packets, melanocytes spread the pigment around so that the keratinocytes, which represent the vast majority of skin cells, also become "pigmented." In response to sunlight, melanocytes produce more melanin, and transfer this melanin more efficiently. That's why we tan.

All normal humans have melanocytes in their skin, and oddly enough, dark-skinned people have roughly the same number of melanocytes as Caucasians. However, the melanocytes of dark-skinned people produce more melanin than do the melanocytes of Caucasians, and their melanin pigment is darker in color.

Skin Cancer and Ultraviolet Light

As organisms evolved on earth, one of the big problems they had to deal with was the possibility of damage to their DNA caused by ultraviolet (UV) radiation. Some organisms responded to this environmental challenge by trying to build a "shell" that would absorb light in the UV range. Most of these guys didn't make it. Other organisms developed sophisticated mechanisms for quickly repairing UV-damaged DNA. And the really smart ones (like humans!) evolved a combination of a UV-absorbing barrier and efficient repair mechanisms.

The function of melanin is to absorb ultraviolet light, thus providing skin cells with a natural "sunscreen" to protect their DNA from the damaging effects of UV radiation. In addition, human cells have sophisticated safeguard systems designed to repair the types of DNA damage that UV light causes—damage which, if not repaired, can lead to cancer. The usefulness of the melanin sunscreen in preventing skin cancer is clear when one considers that the incidence of melanoma in races with darkly pigmented skin is about ten-fold less than in fair-skinned, Scandinavian types. Likewise, the importance of DNA repair in protecting against the effects of UV radiation can be seen vividly in rare humans who have a disease called xeroderma pigmentosum (good words to learn to impress your friends at parties!). Individuals who suffer from xeroderma pigmentosum have a genetic defect in a safeguard system that repairs UV-damaged DNA. Because of this defect, they have a greater than 1,000-fold increased incidence of sun-induced skin cancer.

Although the melanin barrier and cellular DNA repair systems work well to protect against the effects of UV radiation, cumulative, long-term exposure to sun as well as short-term, intense exposure (sunburns) can result in mutations due to unrepaired or incorrectly repaired DNA damage—and these mutations can corrupt cellular control systems. Consequently, exposure to UV light is a major risk factor for all three types of skin cancer.

Because of a growing interest in recreational suntanning, the incidence of skin cancer has increased dramatically. For example, the lifetime melanoma risk for Caucasians in the United States was about one in 1,500 in 1935, when suntanning was not in vogue. Today, it is about one in sixty. Indeed, exposure to UV light now causes more cases of cancer than does cigarette smoking. Although skin cancer can occur (for still-mysterious reasons) in areas of the body that are normally protected from sunlight, the vast majority of skin cancers are associated with exposure to UV radiation. So, as with lung cancer, most skin cancer is totally preventable: Just wear sun block, a hat, and protective clothing.

Genes and Melanoma

In addition to the risk associated with exposure to UV light, there are genes which one can inherit that make it more likely that one will get melanoma. So far, the most important of these susceptibility genes is a tumor suppressor called p16 (how original!). People who inherit a mutated form of the p16 gene frequently suffer from melanoma when they are in their twenties or thirties—an unusually early age for developing this cancer. What is interesting about the p16 protein is that it is part of the same growth-factor pathway that includes pRB, the tumor suppressor that frequently is disabled in lung cancer. In addition to being on the same "relay team," p16 functions in a way that is very similar to pRB. In cells that have not been signaled to proliferate, p16 holds onto another runner in the relay (CDK4) and keeps that protein from passing the baton. When growth factors do call for proliferation, one of the runners on the relay attaches a phosphate group to the p16 protein, causing it to release its grasp on CDK4. That's the normal function of p16. However, if the p16 gene is mutated so that the p16 protein cannot restrain the next runner in the relay, the resulting false start can lead to uncontrolled proliferation.

In some melanoma cells, the gene that specifies CDK4 is mutated so that p16 can't get a grip on it. This mutation also can lead to loss of growth control. The fact that p16, pRB, and CDK4 all function in the same pathway, and that mutations in the genes that specify any one of these proteins can lead to uncontrolled proliferation brings up an important point: Just as the system that controls your furnace has multiple components (thermostat, relay, wire, valve, etc.), every growth factor pathway that has been discovered has multiple components. Consequently, the possibility exists that for each of these components, a mutation may lead to loss of growth control. The growth factor pathway that includes p16, pRB, and CDK4 is a frequent target for mutations associated with many different types of cancer. Indeed, it is likely that some element in the p16/pRB/CDK4 pathway is mutated in one way or another in almost every human cancer.

It is important to note that although control systems frequently have multiple components that can malfunction when mutated, most mutations have little or no effect on the function of a protein. Consequently, for a given protein, only a few, very special mutations will yield a protein that has the shape required to cause a growth-promoting or safeguard system to malfunction. This is a good thing: The small "target size" for cancer-associated mutations helps keep us all from getting cancer all the time.

Screening for Skin Cancer

In contrast to lung cancer, where screening for early stage disease is difficult, screening for skin cancer is quite easy: A visual inspection of a person's skin can detect skin cancers so early in their development that essentially all of them can be cured simply by removing the offending tissue. Basal cell and squamous cell carcinomas appear as pink lesions that persist or recur at the same location, and which frequently have a "pearly," translucent appearance. They usually are easily irritated, becoming red or bloody as the result of slight abrasions.

Melanomas begin life either as moles or, in older people, as "liver spots" (lentigos). Moles and lentigos are both the result of the inappropriate proliferation of melanocytes. As the cells in a liver spot proliferate, they spread laterally across the basement membrane. Consequently, liver spots are generally flat, and they appear dark because of all the melanin that melanocytes contain.

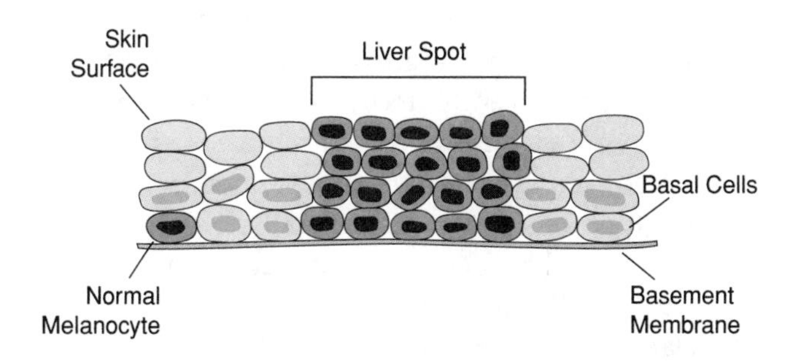

Moles can be either fairly flat or raised, and it is the mounding up of melanocytes, which are proliferating inappropriately, that can lift the mole above the surface of the skin.

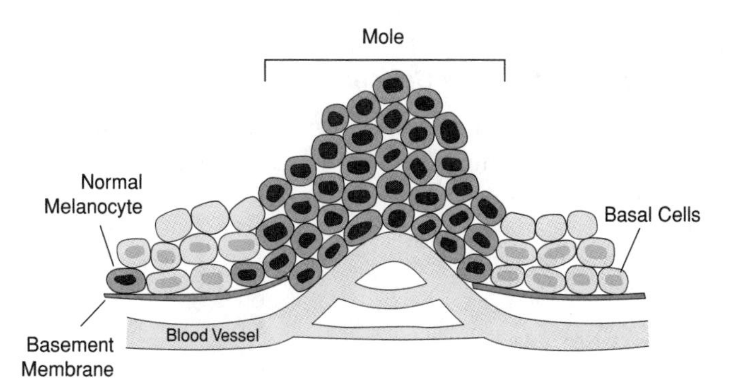

One thing that is important to note here is that the mound of melanocytes actually rises from beneath the basement membrane. This makes sense from the melanocytes' point of view: The blood vessels needed to nourish the melanocytes in the mound lie below the basement membrane. Without access to these vessels, there would be no way the melanocytes could form a mound large enough to raise the level of the skin. From our point of view, however, this means there is impending danger. These melanocytes have already "figured out" how to break through the basement membrane, and how to encourage the growth of new blood vessels within the mound of melanocytes: They are no longer *in situ*. So by the time melanocytes form a mole, the systems within these cells that normally control proliferation, invasion, and angiogenesis have already been corrupted.

Fortunately, most moles are still not dangerous, because additional systems within these cells that control the functions needed for them to metastasize are still intact. The same is true of liver spots. Indeed, "metastatic" systems never will be corrupted in the melanocytes that make up most moles and liver spots. Consequently, moles and liver spots usually don't progress to become metastatic melanomas. The problem, of course, is to differentiate between those moles and lentigos which are benign and those which are cancerous. Fortunately, most moles and liver spots that are progressing toward becoming melanomas have readily recognizable characteristics: They usually are asymmetric, have irregular boarders, are multicolored, and are larger than the eraser on a pencil. Interestingly, it is the flatter moles and the liver spots that are most likely to become malignant melanomas.

Basal cell carcinoma, squamous cell carcinoma, and melanoma all occur most frequently on areas of the body that have been exposed to sunlight—and of course, some of these areas are difficult for a person to view himself. So it is important to have your skin checked regularly by a friend or, better yet, by a physician. And keep a special eye out for moles or liver spots that change shape or color.

Treatments for Skin Cancer

In the early stages, basal cell carcinoma, squamous cell carcinoma, and melanoma all are easily treated by relatively minor surgery. All three cancers are slow-growing, giving ample opportunity for them to be spotted before they metastasize. So in terms of early detection and treatment, skin cancer is very different from lung cancer—which usually goes undetected until it is too late.

If not treated early, melanoma can metastasize, and once it does, the outlook is similar to metastatic lung cancer—very bleak. Unlike most cancers, which have preferred sites to which they travel, malignant melanoma can metastasize to almost any organ in the body. Melanoma metastases generally follow a pattern, however, beginning with skin nodules and then tumors in lymph nodes, lungs, liver, and brain. Later, metastases can also be found in the spleen, intestines, bones, and adrenal glands. Once metastases are discovered in internal organs, the patient usually will survive for less than a year. Chemotherapy, surgery, and radiation therapy can be used to improve the quality of life for some patients, but these treatments rarely prolong life more than a few months. Most patients die from the effects of brain metastases, when the growing tumor crowds out normal tissues, causing the loss of brain function.

THOUGHT QUESTIONS

1. What is the evidence that cigarette smoking greatly increases one's risk of getting lung cancer?
2. Risk factors frequently synergize. What does this mean, and how does it work?
3. Why is there generally a long lag time between exposure to a carcinogen and the onset of cancer?
4. Usually, both copies of a tumor suppressor gene must be mutated before a control system malfunctions. Describe how this works for one of the tumor suppressor genes that is commonly mutated in lung cancer.
5. If you were forced to pick one type of cancer, would you choose to have breast, prostate, lung, or skin cancer? Explain your reasons for making this choice.
6. Compare and contrast lung and skin cancer. How are they similar and how are they different?

Table of Concepts for Lecture 5

Concept	Example
Environmental factors can increase the risk of developing certain cancers.	UV radiation and skin cancer Smoking and lung cancer
Control systems have multiple components, so there usually are several different genes that can be mutated to corrupt a system.	p16 in melanoma RB in lung cancer
Tumor suppressor proteins can sense damaged DNA and halt proliferation until it is repaired.	p53
Tumor suppressor proteins can perform "restraining" functions in growth factor pathways.	pRB in lung cancer p16 in melanoma
Screening procedures can detect certain cancers at an early stage.	Skin cancer
Different cancers usually have favorite sites to which they metastasize.	Lung cancer

Colon Cancer

REVIEW

Lung cancer is an excellent example of a malignancy in which environmental factors can increase your chances of getting cancer. Over 90% of all lung cancers occur in patients who are cigarette smokers, and inhalation of radon gas or its radioactive decay products also is a risk factor for lung cancer. In addition, most cases of a lung cancer called mesothelioma arise in people who have been exposed to asbestos. Importantly, asbestos or radon inhalation plus cigarette smoking is a deadly combination: For those individuals who are exposed to asbestos or radon gas and who also smoke, the synergy of these environmental factors results in an increased probability of contracting lung cancer that far exceeds the sum of the increased risks associated with these environmental factors acting alone.

As in breast and prostate cancer, the primary targets of lung cancer are the epithelial cells that line the ducts of the lungs. However, in contrast to breast and prostate cancer, for which screens exist that can detect cancer at an early stage, most lung cancer is discovered only after it has metastasized to other vital organs, and is impossible to cure. In more than 90% of small cell lung cancers, both copies of the p53 tumor suppressor gene have been mutated. This is a big problem, because the p53 protein is an important player in a safeguard system that senses when DNA has been damaged, and which either stops the injured cell from proliferating or triggers it to die. Once this system has been disabled, the mutation rate in the cell can soar, greatly increasing the likelihood that additional mutations will occur, and that other growth-promoting and safeguard systems will be corrupted.

About 20% of all cancer patients die of cachexia. In this syndrome, patients lose both muscle mass and fat. Cachexia differs from anorexia in that individuals who are anorexic do not lose muscle mass until after all their fat reserves have been exhausted. Little is known about what causes cachexia, and why only certain cancers, for example lung cancer, are associated with this "wasting" syndrome.

Basal cell and squamous cell carcinomas are skin cancers that arise when control systems malfunction in one of the basal epithelial cells that is destined to become a keratinocyte. In contrast, melanoma results when another type of cell in the basal layer of the skin, a melanocyte, loses growth control. For all three types of skin cancer, exposure to UV light is a risk factor, because UV radiation can damage DNA, resulting in mutations that corrupt growth-promoting and safeguard systems. So lung cancer and skin cancer have this feature in common: Environmental factors play a major role in increasing the incidence of these cancers.

Lung cancer and melanoma also have another similarity. Frequently, the same growth-promoting system is defective in both cancers—a system which includes proteins specified by two tumor suppressor genes, p16 and RB. These proteins are members of a "team" which relays a "proliferate" signal from the surface of the cell to the cell nucleus, where genes involved in proliferation are turned on or off in response to this signal. The p16 and pRB tumor suppressor proteins both function by restraining the next "runner" in the relay until they are "tagged" by a "teammate." In each case, the tag is in the form of a phosphate group, which is attached to the tumor suppressor protein, changing its shape, and causing it to release its grasp on the protein to which it is clinging. Both p16 and RB qualify as tumor suppressor genes, because if either is mutated so that the protein it specifies loses its restraint function, the cell will be triggered to proliferate even though no growth factors are present.

Although the p16 and pRB proteins are members of the same team and function in very similar ways, lung cancer cells rarely have p16 mutations and melanomas seldom involve RB mutations. This demonstrates the important point that the mutations required to disrupt cellular control systems often are different in different cell types—even when the same control system is involved. The p16/pRB team also exemplifies the concept that, because all growth-promoting and safeguard systems have multiple components, there usually are several different ways that a given control system can be corrupted. Of course, not every mutation that occurs in a growth-control gene will compromise the system. Indeed, only a small fraction of all the possible mutations result in a protein whose shape is altered in such a way as to disrupt the workings of a growth-control system.

In contrast to lung cancer, which is difficult to detect in its early stages, skin cancer is very easy to identify before it metastasizes. One simply needs to know what skin cancers look like. They are in clear view. However, once either lung or skin cancer has metastasized, the outlook is very similar—extremely bleak.

COLON CANCER

In the United States, the lifetime risk of getting colon cancer is about one in twenty, with most cases occurring in adults who are older than fifty years of age. This year in our country, about 140,000 new cases of colon cancer will be diagnosed, and approximately 55,000 people will die of the disease. So this is certainly one of the biggies. In addition to being interesting from a statistical point of view, colon cancer is also interesting because it is one of the best examples of a cancer that progresses through discrete stages.

The Normal Colon

To understand how colon cancer arises, we first must understand the structure of a normal colon (sometimes called the large bowel). The relative positions of the stomach, small intestine, and colon are shown in this diagram:

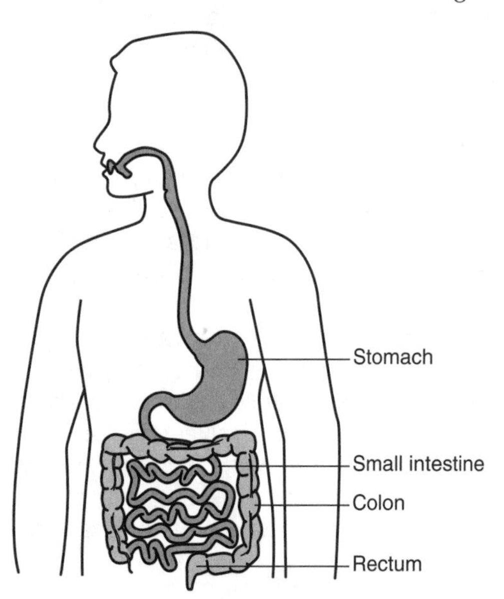

The stomach functions as a storage area for food that is in the process of being digested. In the small intestine, food is broken down into small molecules that can be absorbed to provide nourishment for the body. The colon performs two main tasks. First, water and salts left over after nutrients have been taken up by the small intestine are absorbed back into the body during passage through roughly the first half of the colon. Without this function, we'd all have diarrhea all the time! The second half of the colon, which connects to the rectum, is primarily involved in the storage of fecal matter until it is convenient for it to be released. Cancer almost never arises in the small intestine, and the majority of cancers in the colon occur in the second half. Nobody knows why.

If you were to "take a walk" down the colon (yuk!), it would appear somewhat like a field with lots of gopher holes in it. These gopher holes are called crypts, and they are where the cells that produce mucus are located. The surface of the colon is covered by only a single layer of epithelial cells. This thin covering facilitates absorption of water and salts. Epithelial stem cells are located near the bottom of the crypts, and as these stem cells proliferate, their progeny move upward toward the surface of the colon where they die or are removed by wear and tear. The whole process, from proliferation of a stem cell to death of its progeny, takes less than a week—so cell birth and death go on continuously in the epithelial lining of the colon.

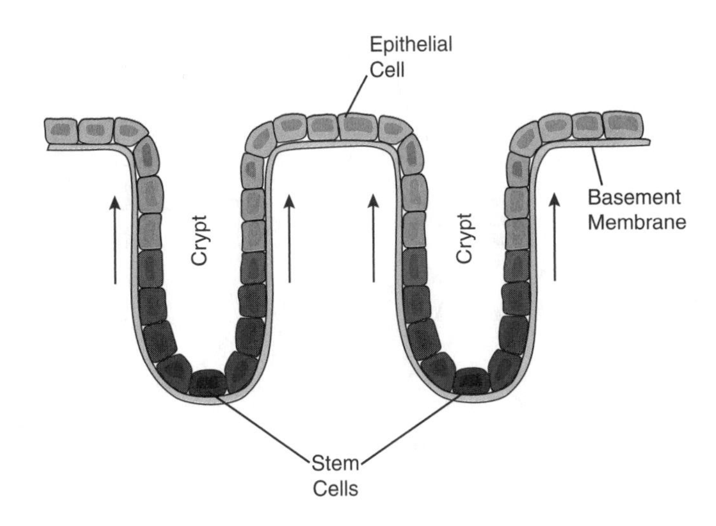

Usually this process is carefully controlled so that the stem cells proliferate just enough to resupply the surface of the colon with fresh epithelial cells. On occasion, however, control systems may be corrupted within one of these epithelial cells, and it may proliferate to form a mass of cells called a polyp, which extends like a finger into the interior of the colon. Most colon cancers evolve from cells which make up these polyps.

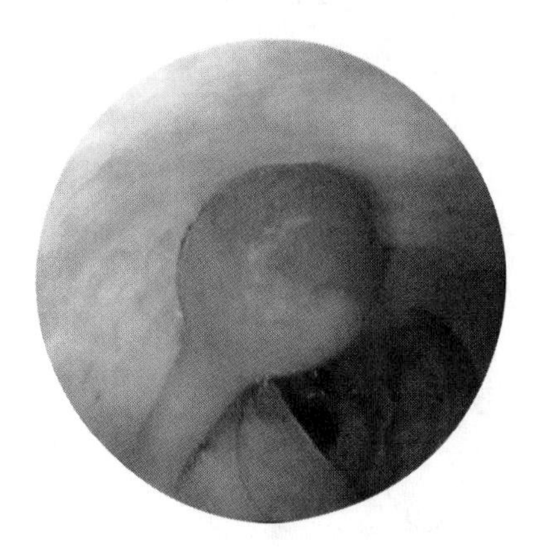

In older people, intestinal polyps are common, and these polyps can become quite large, sometimes reaching a length of half an inch or more. Fortunately, less than 1% of intestinal polyps ever progresses to become a metastatic cancer.

The Progresson to Colon Cancer

Although it is believed that most cancers progress through stages of increasing malignancy as, over time, various control systems are corrupted, colon cancer is one of the cancers in which this step-wise progression is most easily demonstrated. By using a colonoscope, the entire colon can be visualized, and any polyps that are found can be cut off and recovered for further examination. In this way, polyps of increasing size can be probed to see how deranged the architecture of these polyps has become, and the cells within these polyps can be tested for genetic alterations that might lead to control system dysfunction. Recent evidence indicates that in the majority of full-blown colon cancer cells, one proto-oncogene has been converted to an oncogene, and the function of three different tumor suppressor proteins has been lost.

APC Mutations

The earliest genetic defects found in most colon cancers are mutations in both copies of a gene called APC. As a consequence of these mutations, the proteins specified by the APC genes no longer function. Although the picture isn't completely clear yet, it appears that the normal APC protein actually performs two functions which, if lost, can result in a colon epithelial cell beginning to progress toward malignancy.

The first job of a normal APC protein is as part of a growth factor pathway, where it functions to keep another protein in the pathway under control. When we discussed lung cancer, we saw the pRB and p16 tumor suppressor proteins playing this type of "restraint" role. Likewise, in colon cancer, if both copies of the APC gene in an epithelial cell are mutated so that the APC proteins lose their restraint function, the growth factor pathway in which the APC proteins participate gets turned on—even when no growth factors are signaling that proliferation of the epithelial cell is required.

So the first consequence of mutations in the APC genes is that a cell in the colonic epithelium begins to proliferate to form a "nest" of what might be called precancerous cells. The mutated cells in this nest have a growth advantage over their neighboring epithelial cells in that they are less dependent on growth factors to trigger them to proliferate.

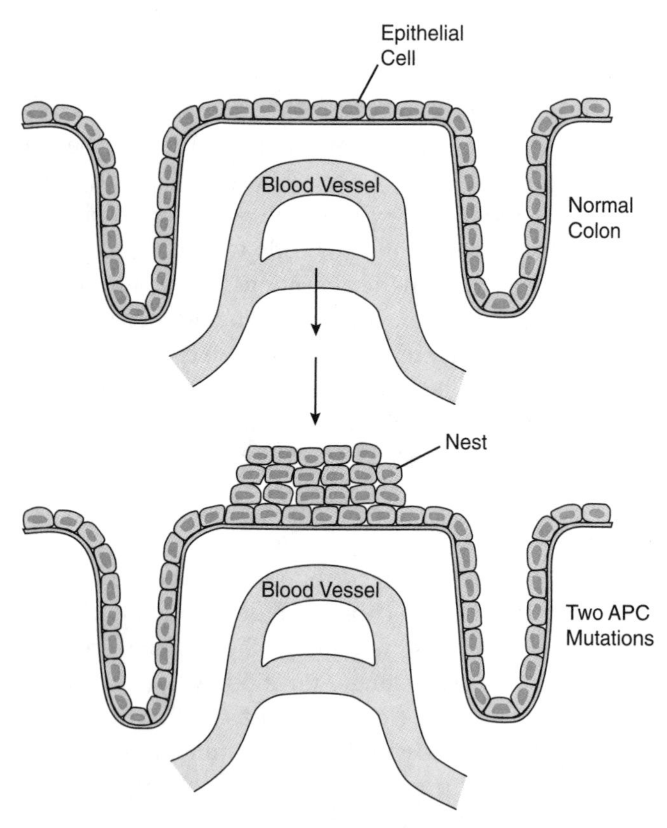

The second job of the normal APC protein is to help ensure that daughter cells each receive a complete set of chromosomes from "mom." When a cell is triggered to proliferate, it must first make a copy of each of its chromosomes (the "volumes" that make up a genetic cookbook). This ensures that when the proliferating cell divides to make two daughter cells, each daughter cell will get a complete set of recipes for all the different proteins it will need. To make this happen, however, a mother cell must solve a tough problem: How to be sure that each daughter receives one and only one copy of a given chromosome once it has been duplicated within the mother cell. The details of how a cell solves this problem are still being worked out, but the overall picture is pretty clear. When a chromosome is copied to produce a pair of chromosomes, a "string" made of protein is attached to each member of the chromosome pair. These strings will be used to direct the chromosomes to the proper daughter cell. To make this happen, the other end of each string is secured to a structure located either at the "north" or "south" pole of the mother cell, so that one member of the chromosome pair is attached to the north pole, and the other member of the pair to the south pole.

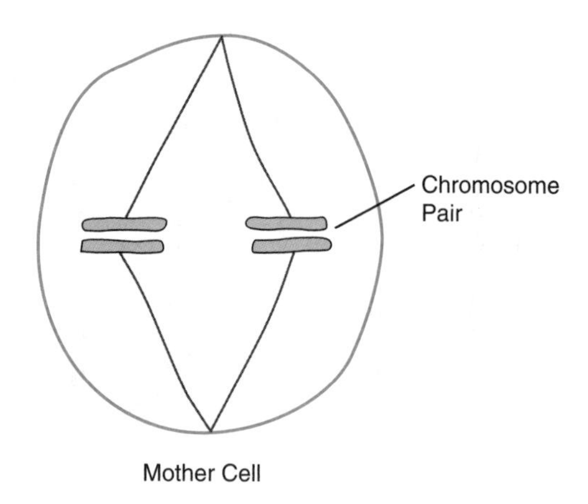

Then, just before the mother cell divides to make two daughter cells, the strings are reeled in so that one copy of each chromosome is pulled to the pole that will become part of one of the daughter cells.

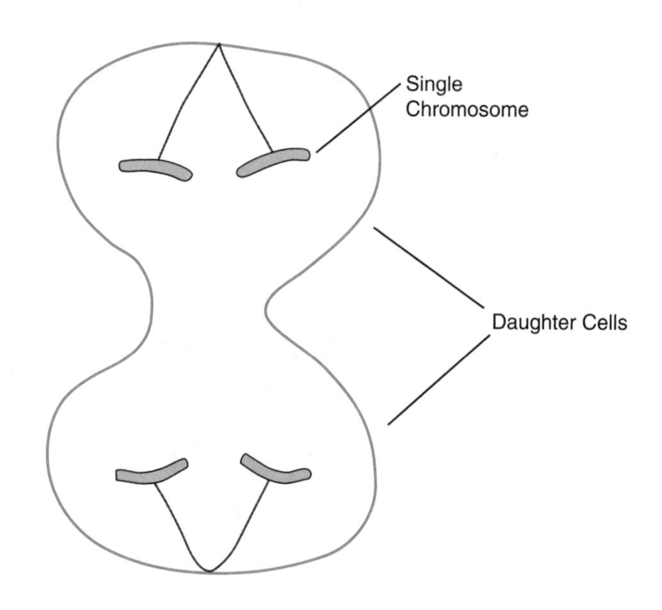

The end result of all this pulling and tugging is that each daughter cell gets one and only one copy of each duplicated chromosome. This mechanism is intended to ensure that each of our cells ends up with a complete copy of the genetic cookbook—and no extra volumes (chromosomes). However, occasionally mistakes are made in this process. For example, if one of the strings fails to hook up to one of the duplicated chromosomes, that chromosome will be "undirected." As a result, the chromosome will be dealt out randomly to one or the other of the daughter cells. When this happens, one of the daughter cells may end up getting both copies of that chromosome and the other daughter cell may receive none. As you can imagine, a cell with too many or too few chromosomes will have severe problems regulating its activities.

Another difficulty that can arise as duplicated chromosomes are divided up is that the string from the north pole and the string from the south pole may both attach to the same chromosome. When this mistake occurs, and the cell starts to divide, the chromosome can be ripped apart, with one piece of the chromosome ending up in one daughter cell, and the other piece in the other daughter cell. This isn't cool either. Indeed, one of the hallmarks of colon cancer cells is that they have "unstable" chromosomes which are constantly being broken, lost, or gained.

Biologists have now shown that in cells that lack the APC protein, many more mistakes are made during chromosome distribution than in normal cells, indicating that APC plays some role in ensuring the accuracy of this process. Consequently, colon cells that have suffered mutations which knock out the function of their APC proteins actually have two strikes against them in terms of becoming cancer cells. First, they have a growth advantage (decreased dependence on growth factors) which allows them to proliferate when normal colon epithelial cells wouldn't, and second, they mutate more rapidly than normal cells, due to faulty chromosome distribution. Because in both cases it is the loss of function of the APC protein which contributes to a cell becoming cancerous, the APC gene is regarded as a tumor suppressor.

Fortunately, very few cells that have suffered APC mutations go on to become metastatic cancer cells. There are two reasons for this. First, there are multiple growth-promoting systems in every cell, and to decide whether to proliferate, a cell evaluates the signals it receives from all of these systems. So if the growth-promoting system in which the APC protein plays a role is the only growth-promoting system that is corrupted, the cell may proliferate somewhat more rapidly than its normal neighbors, but probably not a whole lot more rapidly. Said another way, the APC mutation may confer a growth advantage on the cell, but not a big advantage.

The second reason only a small fraction of cells with mutations in their APC genes go on to become cancerous is that there are safeguard systems within each cell that monitor not only unscheduled proliferation, but also faulty chromosome distribution. Thankfully, these safeguard systems take care of most wannabe cancer cells that have APC mutations.

KRAS Mutations

Sometimes, however, additional mutations may occur that can cause one of the cells within a nest of cells with APC mutations to progress to the next stage in colon cancer development: polyp formation. In about half of all polyps, this second mutation occurs in a gene called KRAS.

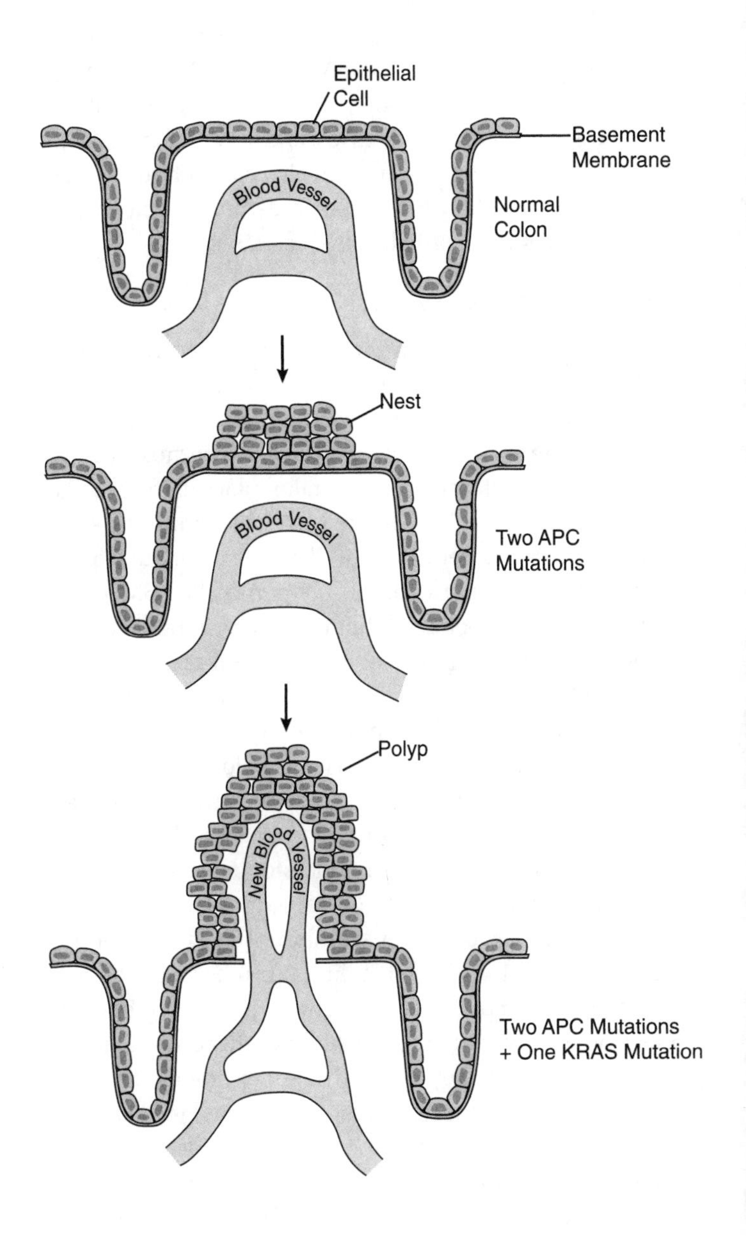

The RAS proteins are actually a family of proteins whose members participate in several different growth factor pathways. The genes that specify RAS proteins are recognized as proto-oncogenes, because the RAS genes can suffer mutations which cause the RAS proteins to be "always on." When this happens, the mutated RAS proteins convey the "let's proliferate!" signal regardless of whether the appropriate growth factors are present. As is usual with proto-oncogenes, only one of the two copies of the KRAS gene need be mutated to signal unregulated proliferation.

Once an epithelial cell has suffered mutations in both copies of its APC genes—plus a mutation in one of its KRAS genes—the cell can proliferate even more rapidly and can form a polyp. In essence, mutations in the two growth-promoting systems (the APC and KRAS pathways) work together to cause more, unscheduled proliferation than the APC mutation alone was able to do. Interestingly, about half of all polyps do not contain cells with the KRAS mutation, so it is likely that other, unidentified proto-oncogenes can substitute for KRAS in these cases.

SMAD and p53 Mutations

At this point, the cells in the growing polyp have suffered at least three mutations, and the polyp continues to grow in size, albeit slowly. So far, the polyp still qualifies for the designation of "precancerous," because additional genetic alteration must take place before cells from the polyp can invade the blood or lymphatic system and metastasize. In about 75% of metastatic colon cancers, another tumor suppressor gene called SMAD is mutated, and it is believed that mutations in both copies of the SMAD gene are involved in the cell acquiring the potential to metastasize. Exactly how this mutation confers metastatic potential is not known.

Roughly 75% of metastatic colon cancers also have mutations in both copies of the p53 tumor suppressor gene, but again, the role that the mutant p53 proteins play in metastasis is not well understood. It has been hypothesized that, because the safeguard system that includes the p53 protein causes normal cells to commit suicide when things "get a little strange," mutations which knock out p53 function may keep metastatic colon cells from throwing in the towel if growth conditions are suboptimal when they reach their new homes in other parts of the body. But that's just an educated guess.

So the current thinking is that multiple mutations drive colon cells to progress through stages in which they become increasingly more "cancer-like." If the sequence of events I have just described is correct, at least seven mutations are required for a normal epithelial cell in the colon to progress to become part of a metastatic cancer: two mutations in APC genes, one mutation in a KRAS gene, two mutations in SMAD genes, and two mutations in p53 genes. Fortunately, this whole process takes quite some time: Usually twenty to forty years elapse from the moment an epithelial cell suffers mutations in both APC genes until cells from a polyp metastasize.

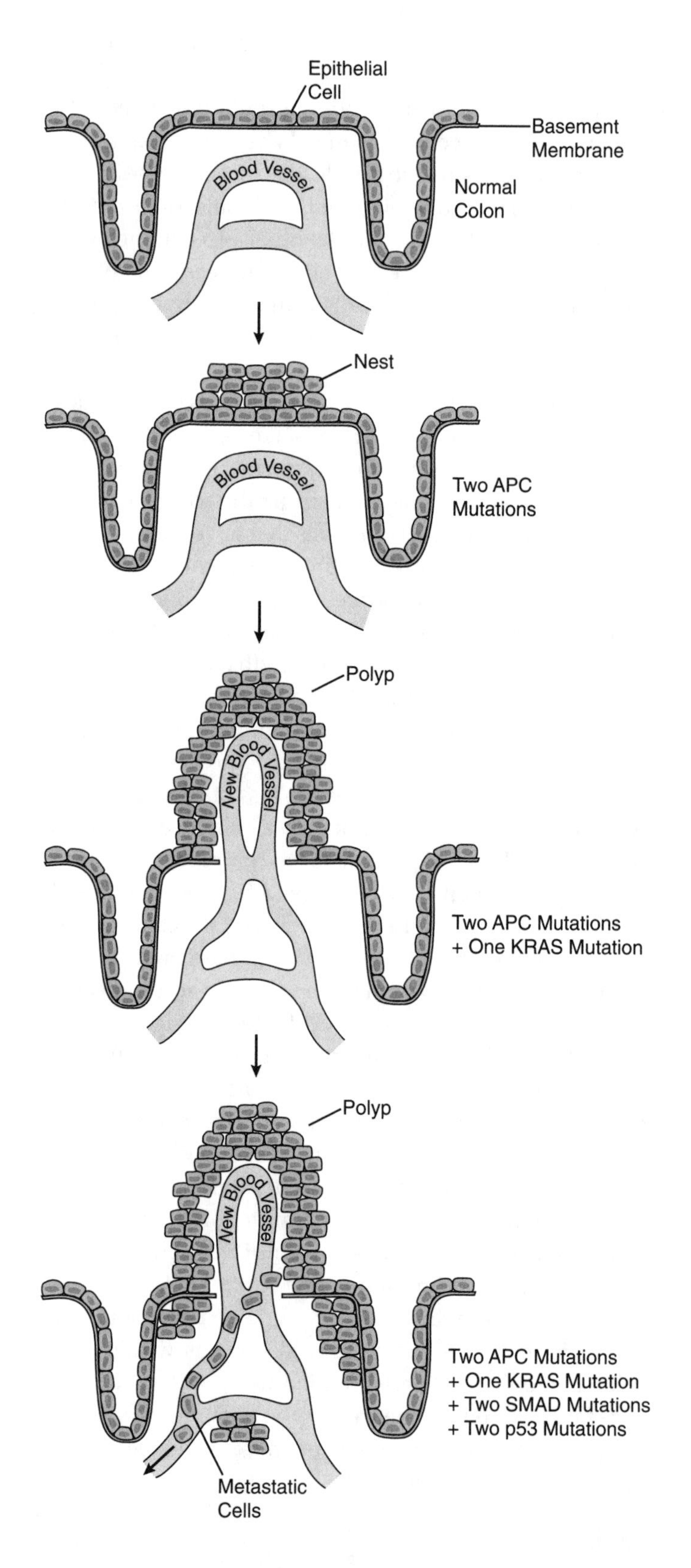

Immortal Stem Cells

You may be wondering how colon epithelial cells, which usually have a lifetime measured in days, manage to stick around for forty years to accumulate all these mutations. It's a good question. Many biologists believe that it is actually the stem cells that reside in the crypts which suffer the initial mutations that lead to colon cancer. Most cells in the body are "mortal," having a limited potential for proliferation. One reason for this is that DNA is progressively lost from the ends of chromosomes as they are copied. Fortunately, there is enough "extra" DNA (called telomeric DNA) at the ends of "young chromosomes" so that the DNA in an average cell can be copied about seventy times before the extra DNA is used up, genes are lost, and the cell no longer can function. In contrast to most cells, which can't keep proliferating indefinitely, stem cells are "immortal." These special cells produce an enzyme called telomerase that adds DNA to the ends of chromosomes, replenishing the DNA that is nibbled away as chromosomes are copied. Because of their immortality, stem cells are likely candidates for the cells in which cancer-causing mutations accumulate over a long period of time.

Diet and Colon Cancer

Colon cancer also provides an excellent example of the difficulty in assessing the connection between diet and cancer. It's pretty clear that a person's diet can have an important effect on the probability that an individual will get colon cancer. For example, the death rate from colon cancer varies over about a forty-fold range in different parts of the world. Although some of these differences may be genetic, environment and diet clearly play large roles, since when people move from one area to another, their death rates from colon cancer begin to resemble those of their new neighbors.

What dietary factors might cause these dramatic variations in colon cancer incidence? Most epidemiologists would agree that a person who is not overweight, and who consumes a diet that is consistently low in red meat and animal fat and is high in fresh fruits and vegetables has a reduced risk of getting colon cancer relative to those in the general population. However, this statement probably belongs in the "well, duh" category: If you eat sensibly, you are more likely to be healthy.

The problem in understanding the relationship between diet and colon cancer arises when one tries to identify individual dietary components that increase or decrease the risk of getting this cancer. Most large-scale studies are "retrospective," in that they ask people to describe their diet and activities over some period in the past. However, most people eat a diet that includes many different elements, and over a 20-year period, the mixture of these elements is certain to change. After all, who eats the same foods every day for twenty years? Further, some elements of a diet may have nothing directly to do with cancer susceptibility, but may just be indicative of a certain lifestyle. For example, people who drink wine at dinner every night may have lower rates of colon cancer. However, the wine itself may have nothing to do with this protection: Drinking wine may simply be a "marker" for people who are more affluent and therefore may have access to healthier foods and better health care (beer drinkers please don't take offense—this is just a hypothetical example!).

What you'd really like to do to probe the connection between diet and cancer is to take two groups of people with similar lifestyles, and ask the people in one group to change something in their diet. For example, you could ask one group to eat Brussel's sprouts three times a week. Then, at some future date, you could compare the cancer incidence in the two groups to see if eating a ton of

Relative Colon Cancer Death Rates

Location	Approximate Relative Death Rate
Southern Africa	1.0
Central America	1.5
Australia/New Zealand	3
Japan	13
Western Europe	31
North America	34
China	42

Data adapted from Pisani, P., et al., *International Journal of Cancer* (1999) Vol. 83, pp. 18–29.

Brussel's sprouts might decrease the risk of getting colon cancer—all other things being equal between the two groups. However, there's a problem with using this strategy to study diet and colon cancer. Usually, it takes many years for colon cancer to develop, so ideally, you'd like this study to last a decade or two. But who's going to eat Brussel's sprouts three time a week for ten years? Certainly not me! I think I'd rather die of colon cancer.

To circumvent the problem of continuing a trial for many years, researchers typically look for "indicators" that may tell them much earlier whether a certain dietary change is helpful in preventing cancer. Because colon cancer usually progresses from small polyps to larger polyps to metastatic cancer, the presence or growth of polyps might be useful as an early indicator. However, checking for polyps requires a procedure like a colonoscopy that is not something you would do on a Saturday night for fun. So finding people who have no previous history of colon abnormalities, and who are willing to participate in a study that involves a colonoscopy every year for even four or five years is rather difficult. Indeed, most studies of this type enroll people who already have had colon polyps, and even then, the number of people who are willing to participate is usually small—too small to say definitively whether a certain dietary item makes a real difference.

Because of the difficulty in studying the connection between diet and colon cancer, no individual dietary component has been shown conclusively to protect a person from or predispose a person to this type of cancer. Diet does seem to matter, but just what the "ideal" cancer-protective diet might be isn't known.

Screening for Colon Cancer

As my gastroenterologist, Paul Hendrix, pointed out to me, the most common symptom of colon cancer is no symptom at all. This means that by the time you "feel" that something is wrong, it is frequently too late to cure the cancer that is growing in there. We have already talked about three types of cancer for which screens can reveal the presence of a tumor before any symptoms appear. For breast cancer, there is the mammogram; for prostate cancer, the PSA test; and for skin cancer, examination of the surface of the skin by someone trained to recognize precancerous lesions. Fortunately, there are screens that can detect colon cancer at an early stage, although the best of them is more "invasive" (i.e., uncomfortable) than the other screens we have discussed.

The Fecal Occult Blood Test

One screen for colon cancer, the fecal occult blood test, is designed to detect blood in the feces. Large polyps in the colon can be abraded by the passage of fecal matter, causing them to bleed. Usually, the amount of blood is too small to be observed visually in the stool. However, tiny amounts of blood can be detected if a smear from a stool sample is collected and sent to a laboratory. Trials involving hundreds of thousands of people have demonstrated that if several stools are sampled (e.g., on successive days), and if the fecal occult blood test is performed yearly, deaths from colon cancer could be decreased by about a third. The reason this test is not more useful is that many polyps don't bleed—or they may bleed only occasionally—so a procedure that tests for blood in the stool only once a year may miss a lot.

Flexible Sigmoidoscopy

A second way to screen for colon cancer is to use an instrument called a flexible sigmoidoscope. In this procedure, a tube about as big around as your thumb (if you have a small thumb) is inserted through the rectum up into the colon. Using fiber optics, this procedure allows roughly the last third of the colon to be inspected for polyps. The flexible sigmoidoscope can also be used to take biopsies of suspicious areas, and to remove small polyps.

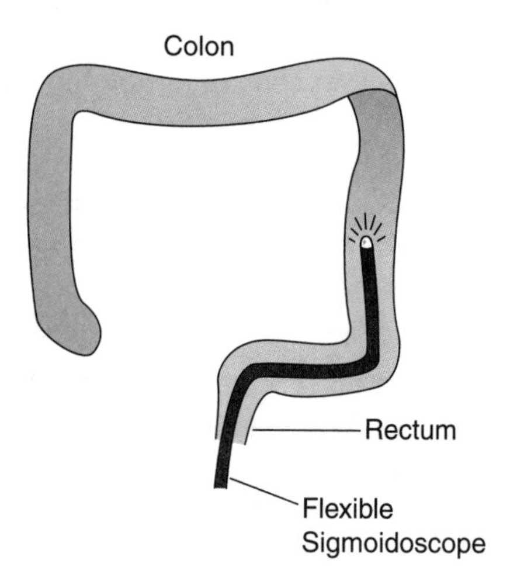

A large-scale study is now underway to determine the usefulness of this procedure as a screen. However, based on smaller, less-well-controlled studies, it appears that having your colon examined every few years using a flexible sigmoidoscope can reduce your chances of dying of colon cancer by about 80%. Flexible sigmoidoscopy has the advantage that although it is uncomfortable, it is a relatively simple procedure that can be carried out in a doctor's office without an anesthetic. The main disadvantage of flexible sigmoidoscopy is that only part of the colon can be viewed. Getting the scope around those two big bends at the top would really be uncomfortable for the patient. Fortunately, about 70% of colon cancers arise in the area that can be surveyed using a flexible sigmoidoscope. So if you have to look at only part of the colon, this is certainly the part to view. However, because a lot can be missed, this procedure has been likened to having a mammogram of only one breast.

Colonoscopy

A third screen for colon cancer is the colonoscopy, in which the entire colon is inspected. This procedure is usually done under mild sedation, because the tube that is inserted has to be coaxed around a total of four bends.

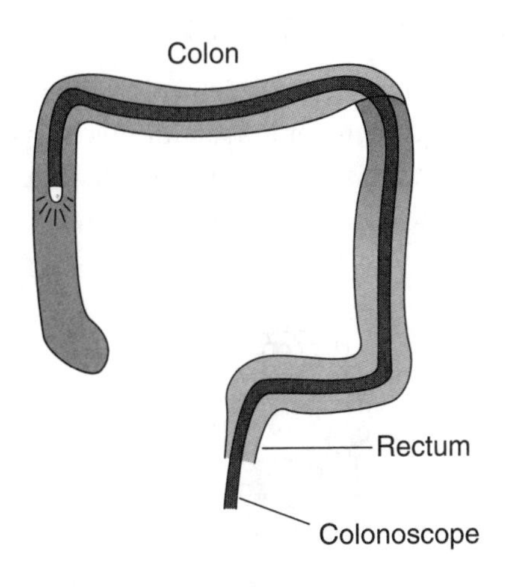

A colonoscopy is truly an amazing procedure which requires great "artistry" on the part of the gastroenterologist who performs it. The plastic tube which is inserted through the rectum is only about the diameter of your little finger, yet this instrument not only allows the doctor to see every inch of the colon, but, by using special "tools" that are passed through the plastic tube, he can cut off and retrieve any polyps he may find in there. Because these doctors work at a distance from the objects they are trying to manipulate, and because their view of these objects is fairly limited, I suspect gastroenterologists would also make good puppeteers. The tool they use which I find most fascinating is the "snare." This is a wire loop which can be positioned around the stalk of a polyp.

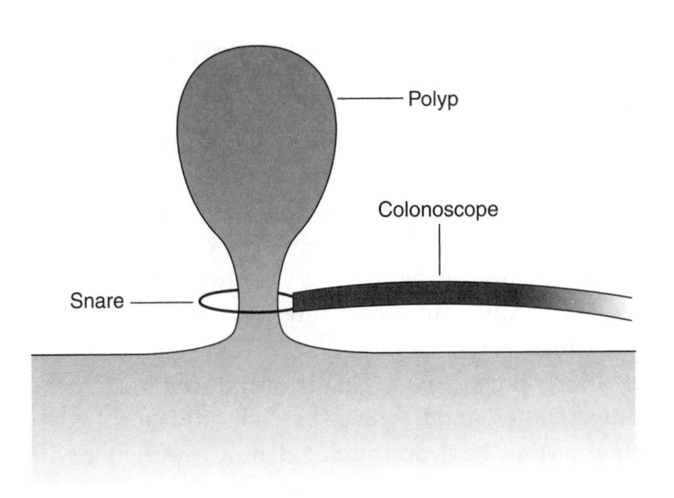

After the gastroenterologist has succeeded in lassoing a polyp (these guys probably also make good cowboys!), he then sends an electric current through the wire, and tightens the noose around the polyp. The electric current heats the wire and helps it cut through the stalk of the polyp. In addition, the hot wire also sears the base of the polyp to prevent bleeding. Once the polyp has been clipped off, suction is applied through a tube that runs inside the outer tube, and the polyp can either be sucked through the scope and recovered, or for larger polyps, suction can be used to hold onto the polyp while the scope is retracted and the polyp is retrieved. The process of retrieval is important, because the cells of the polyp will be examined by a pathologist to determine whether the polyp is benign or cancerous.

No studies have been completed which evaluate the effectiveness of colonoscopies in screening for colon cancer. However, most physicians believe that as a screening tool, colonoscopy is more effective than flexible sigmoidoscopy, mainly because this technique lets the gastroenterologist examine the entire colon. Indeed, if polyps are detected using a flexible sigmoidoscope, the next step is usually a colonoscopy to try to get any polyps that may be "hiding" further up the colon. Since about 30% of people over the age of fifty will have one or more polyps in their colons, the argument could be made that if you are fifty or older, you might just as well go for the whole banana—the colonoscope.

Before we get off this subject, I want to say one further word about early cancer detection. Cancer that is detected early, frequently can be cured. Cancer that has metastasized, almost never can be cured. So it makes sense to me that if screens are available which can detect cancer at an early stage, we should take advantage of them.

Staging and Treatment of Colon Cancer

The usual treatment for colon cancer is to remove the offending tissue. Exactly how much tissue to remove and whether additional treatments are required will depend on the "stage" of the cancer at the time of detection. Colon cancer is an excellent example of how tumor staging can be used to determine the most effective treatment and the likely outcome for a patient with cancer. The most frequently used staging system for colon cancer was originally introduced by a doctor named Dukes, and has since been slightly modified by others as they have accumulated more experience with colon cancer. This system takes into account the depth below the surface of the colon to which the cancer has penetrated; whether tumor cells can be detected in lymph nodes that drain lymph (and the cancer cells this lymph may contain) from the site of the tumor; and whether metastases can be detected at other locations in the body. For example, a polyp that extends only a short distance below the basement membrane almost always can be cured simply by "clipping it off." Such a lesion is classified as Dukes stage A. At the other end of the spectrum is a colon cancer that has already metastasized to a distant site: a Dukes stage D tumor. These patients usually survive for less than one year after their colon cancer is diagnosed.

For patients with Dukes stage A or B1 colon cancers, the probability that removal of the cancerous tissue will produce a cure is so great that no additional treatment is recommended. Although radiation or chemotherapy might increase the cure rate of stage B1 cancers marginally, the risks associated with these additional treatments outweigh the possible gains.

Radiation therapy or chemotherapy sometimes is used to "mop up" any cancer cells that may have escaped the surgeon's knife. Patients who are good candidates for this type of "adjuvant" therapy have stage B2 or C tumors. Another use for radiation or chemotherapy is to try to extend the patient's life. In this situation, these treatments are not expected to cure the patient, but rather to help him survive longer. Patients with stage C tumors are candidates for this type of therapy. Radiation therapy and chemotherapy can sometimes be used to destroy enough of the primary tumor to make surgical removal easier (or possible). This is most helpful for large, stage C or D colon cancers. Finally, for colon cancer in the advanced stages, radiation or chemotherapy is sometimes employed to shrink metastases in order to make the patient more comfortable. These examples illustrate the point that categorizing tumors according to the stage to which they have progressed can be very helpful in determining which therapy is appropriate for a given patient.

Colon Cancer Metastases

The most common site to which colon cancers metastasize is the liver. The major reason for this is ease of transport. The colon is an organ in which fluids that are passing through the digestive tract are absorbed by blood vessels that lie just below the epithelial surface. This blood is then collected and piped through the liver on its way back to the heart. The reason for this plumbing scheme is that the colon also contains toxins which have been ingested or which were produced during the digestive process—and these toxins can also be absorbed by the blood vessels around the colon. By routing

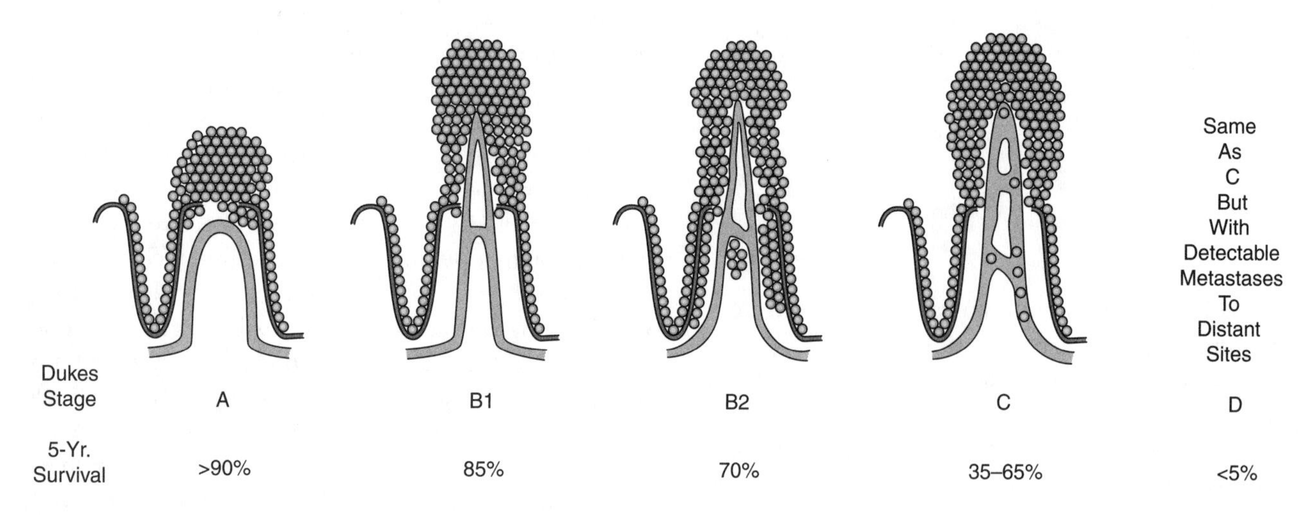

this blood first through the liver, the major "detox center" in the body, these toxins can be rendered harmless (or at least less harmful) before the blood goes into the general circulation. Unfortunately, this same plumbing system also insures that colon cancer cells which have invaded blood vessels around the colon will go directly to the liver where they may proliferate to form a metastatic tumor.

Limitations of Radiation Therapy and Chemotherapy

If radiation and chemotherapy can, in some cases, reduce the probability that a primary tumor will metastasize, and can shrink tumors in others, why can't these two treatments be used to cure all cancers? The reason is that most cancer cells mutate so rapidly that a tumor usually contains some mutant cells that are resistant to radiation therapy or chemotherapy. In this respect, it is important to keep in mind that a tumor is actually a collection of cells with many different mutations. Because a tumor can easily contain 100 billion cells, this collection of mutant cells can be huge.

Biologists are discovering more and more ways that tumor cells can mutate to resist anti-cancer therapies. Many of these therapies are designed to injure cancer cells so seriously that they commit suicide. When we discussed follicular lymphoma, we noted that a protein called bcl-2 could prevent cells from committing suicide if it were expressed at higher-than-normal levels. Solid tumors (e.g., colon cancers) may also contain cells that have mutated to produce high levels of the bcl-2 protein, making these cancer cells resistant to killing by radiation and chemotherapy. In fact, the "suicide system" within cells is so complex that there are many different genes which, if mutated, can result in this system being disabled. The bcl-2 gene is just one example.

There are other mutations that can render tumor cells resistant to anti-cancer therapies. For example, some mutant cells resist chemotherapeutic agents simply by blocking entry of these drugs into the cell. Many drugs (e.g., methotrexate and cisplatin) enter cells by using systems whose normal function is to transport certain nutrients into the cell. Cancer cells can become resistant to such drugs if they sustain mutations which shut down these transport systems either partially or completely. When this happens, the chemotherapeutic drug may not be able to enter the cell at all, or may enter so inefficiently that it never reaches an effective concentration within the cell.

Once inside a cell, some drugs (e.g., 5-FU) must be modified by cellular enzymes before they can be effective. If the genes that specify these modifying enzymes are mutated so that the enzymes no longer function, the drug will remain inactive (because it has not been modified), and the cell will be resistant to this chemotherapeutic agent.

Radiation therapy and many anti-cancer drugs (e.g., cyclophosphamide and melphalan) work by damaging DNA. However, cells have systems that can repair damaged DNA, and if mutations occur which make these repair systems function more efficiently, mutant cells can resist the effects of DNA-damaging agents. This tactic is very interesting in view of the fact that we usually think of cancer cells as having mutations which disable systems designed to repair DNA. The trick here is that every cell has multiple DNA repair systems, and each system specializes in fixing certain types of DNA damage (e.g., double-strand breaks). So a cancer cell can win big if it has sustained mutations which increase the efficiency of those systems which repair the particular type of damage inflicted by the chemotherapeutic agent—as well as mutations which disable other DNA damage repair systems, resulting in a rapid mutation rate. In this way, the cancer cell can have its cake (mutate rapidly) and eat it too (resist chemotherapy)!

Some chemotherapeutic drugs (e.g., methotrexate) kill cancer cells by inactivating cellular enzymes that are critical for cell survival. However, if cancer cells mutate so that the target enzyme is over-produced, there may not be a high enough concentration of the drug within the cell to disable all the target enzymes, allowing the mutant cell to survive. Cancer cells also can resist destruction by some anti-cancer drugs (e.g., vinblastine and doxorubicin) by pumping these drugs right back out of the cell before they can do any damage! This "export" system is used by normal cells to protect them from toxins that are generated as everyday byproducts of the chemical reactions that go on inside a cell. However, if due to mutations, these pumps get cranked up to high levels in a cancer cell, they can render anti-cancer drugs ineffective. Because these pumps are designed to protect normal cells against a wide range of toxins, cancer cells with over-active pumps frequently are simultaneously resistant to the effects of multiple chemotherapeutic drugs ("multidrug resistance").

So there are many ways cancer cells can resist the effects of radiation therapy and chemotherapeutic drugs. The important concept here is that because most cancer cells mutate rapidly, the likelihood is great that somewhere in the billions of cells in a tumor there will be cells with the right mutations to resist killing by radiation therapy or by chemotherapeutic agents. Consequently, although most of the cells that make up a tumor may be destroyed by anti-cancer treatments, resistant cells frequently remain which can proliferate to "re-stock" the tumor.

It is important to understand that cancer cells don't mutate in response to anti-cancer therapies. The mutations which lead to resistance actually occur before treatments begin. It's just that the number of different mutations represented in the "collection" of cells that make up a tumor is so large that there are likely to be some cells which can resist almost any treatment.

THOUGHT QUESTIONS

1. Colon cancer progresses through various "stages." Explain what this means and give examples of "genetic events" that cause colon cancer cells to progress from one stage to the next.
2. Why are stem cells good candidates for cells that eventually become cancerous?
3. There are three tests that can be used to detect colon cancer in its early stages. What features of colon cancer make early detection possible?
4. Why are radiation therapy and chemotherapy frequently unable to cure cancer?
5. What are some of the mechanisms that cancer cells use to resist anti-cancer therapies?
6. Do cancer cells mutate in response to anti-cancer therapies?

Table of Concepts for Lecture 6

Concept	Example
Many cancers probably arise due to mutations that accumulate in immortal stem cells.	Colon cancer
Multiple control systems must be corrupted to produce a cancer cell.	Colon cancer
The same cancer can result from different mutations in different control system genes. Each cancer is a "family" of diseases.	Colon cancer
Oncoproteins can short-circuit growth factor pathways.	KRAS in colon cancer
Tumor suppressor proteins can perform "restraining" functions in growth factor pathways.	APC protein in colon cancer
Tumor suppressor proteins can oversee the orderly distribution of chromosomes.	APC protein in colon cancer
Some cancers progress in well-defined stages.	Colon cancer
Screening procedures can detect certain cancers at an early stage.	Colon cancer
Different cancers usually have favorite sites to which they metastasize.	Colon cancer
Cancers frequently contain mutant cells that can resist anti-cancer treatments.	Colon cancer

Cancer of the Cervix and the Liver

REVIEW

Colon cancer provides a clear example of a cancer that progresses through stages of increasing malignancy as more and more control systems are corrupted. Mutations in both copies of the APC tumor suppressor gene result in cells that have a growth advantage over their neighbors, and which also are prone to make mistakes as they pass out chromosomes to their daughter cells. Because of their growth advantage, cells with this mutation proliferate to form a nest of precancerous cells. Moreover, disruption of the system that controls chromosome distribution increases the mutation rate of the cells in this nest, and sets the stage for corruption of additional control systems.

In about half of all colon cancers, the next system to be corrupted is a growth-promoting system that includes the KRAS protein. If one copy of the KRAS proto-oncogene is mutated so that the protein it specifies is "always on," the KRAS growth-promoting system drives the mutant cell to proliferate more rapidly, converting the nest of cells into a growing polyp.

Fortunately, the vast majority of polyps will never develop into a metastatic cancer. However, if a cell within a polyp suffers additional mutations in two other tumor suppressor genes, the cell and its progeny can metastasize to distant parts of the body—the final stage in colon cancer development. This progression of colon epithelial cells from normal to metastatic usually takes several decades, and is believed to require mutations in at least seven different genes that make up growth-promoting and safeguard systems. Because incipient colon cancer cells must "stick around" long enough to accumulate all these mutations, colon cancer probably arises when mutations occur in one of the "immortal" epithelial stem cells.

Colon cells progress from normal to metastatic in steps that are fairly well defined, so a determination of the stage which a colon cancer has reached at the time of diagnosis can be quite helpful in determining both the appropriate treatment and the patient's prognosis. This "staging" takes into account how deeply the cancer has invaded the tissues surrounding the colon, whether cancer cells can be detected in nearby lymph nodes, and whether the cancer has metastasized to other parts of the body.

Colon cancer is one cancer in which diet clearly plays a role. However, it has been impossible to identify any individual dietary component which can protect against colon cancer or that can predispose a person to this malignancy.

Colon cancer is one of the five cancer types for which screens exist that can detect tumors during the early stages of their development—at a time when they usually can be cured. Three screening procedures are available for the early detection of colon cancer: occult blood testing, flexible sigmoidoscopy, and colonoscopy.

Although radiation therapy and chemotherapy can be useful in treating cancer, the success of both therapies in curing cancer frequently is limited by the ability of some cells within a tumor to resist these treatments. Most cancer cells mutate so rapidly that included in the billions of cells which make up a tumor, there are likely to be cells that have suffered mutations which enable them to resist anti-cancer treatments. These mutant cells are present in the tumor before the anti-cancer treatments are administered, so cancer cells don't mutate in response to the treatments.

> Mutations can render cells resistant to anti-cancer treatments in a number of different ways. These include blocking entry of a drug into the cell, strengthening systems that repair damaged DNA, disabling cellular enzymes required to activate chemotherapeutic drugs, and even pumping drugs back out of the cell before they can do damage.

CANCER OF THE CERVIX AND THE LIVER

In this lecture, we will examine another unlikely pairing: liver and cervical cancer. What these cancers have in common is that viral infections are a major risk factor for both types of cancer.

Cervical Cancer

You may have heard that viruses can cause human cancer. This is not true. There is no virus that causes human cancer. What can be said about viruses and cancer is that there are a few viral infections which can increase the risk of getting certain cancers. So just as cigarette smoking is a risk factor for lung cancer, infection with certain viruses can be a risk factor for particular cancers. The connection between infection with human papilloma virus (HPV) and cervical cancer is an excellent example of this concept.

Human Papilloma Virus

Human papilloma viruses comprise a family that has about a hundred, slightly different members. Each type of HPV infects specific regions of the human body, and some are spread by activities as casual as shaking hands or walking on the deck of a swimming pool. Other types of human papilloma virus are spread by intimate physical contact, and some of these can play a role in cervical cancer.

Each human papilloma virus is essentially a small piece of DNA wrapped in a protective coat of protein. In contrast to a human cell, which contains about 35,000 different genes, the "genetic cookbook" of HPV is limited to only eight genes. Because it carries such a limited amount of genetic information, HPV (like every other virus) must live as a parasite within human cells. When HPV enters a cell, it sheds its protein coat, and uses the machinery of the cell to make many copies of its viral DNA and to produce more coat proteins. Newly minted DNA molecues are then cloaked in protein and leave the cell to infect other, nearby cells. Usually, a single infected cell can produce thousands of new viruses, so an infected cell literally becomes a "virus factory."

HPV is extremely picky about the cells it infects and the conditions under which it will reproduce. The targets of an HPV infection are the basal stem cells which are located beneath our skin and all our mucosal surfaces. These basal stem cells proliferate on demand to replace epithelial cells as they are lost or damaged. Some of the progeny of these stem cells remain attached to the basement membrane and become stem cells themselves, whereas others are pushed upward toward the surface. After they have become disengaged from the basement membrane, the epithelial cells mature and stop proliferating.

As they rise toward the surface of the skin, propelled by continued basal cell proliferation, the maturing epithelial cells of the skin dedicate themselves to the production of keratin proteins. Nearing the surface, the cells eventually die, becoming flattened bags of keratin, which function as overlapping "shingles" to provide protection against the outside environment. As these dry shingles flake off the skin surface due to normal wear and tear, they are replaced by cells rising from below. I mention skin here because, although we are discussing cancer of the cervix, HPV is usually thought of as a "wart virus." Indeed, infection with some types of human papilloma virus can cause the common warts we all have "enjoyed" at one time or another.

For the mucosal surface that lines the female reproductive tract, the story is a bit different from that of skin. As daughter cells are pushed higher and higher by proliferating basal cells, they flatten out, but they do not produce large amounts of keratin proteins—and they remain alive. Consequently, the cells at the top of the stack (the squamous cells) are more like moist pancakes than dry shingles. Keratinized shingles make a fine covering for skin, but a dry, flaky vagina would not be so good. As the squamous cells in the upper layer slough off, they are replaced by others rising from below.

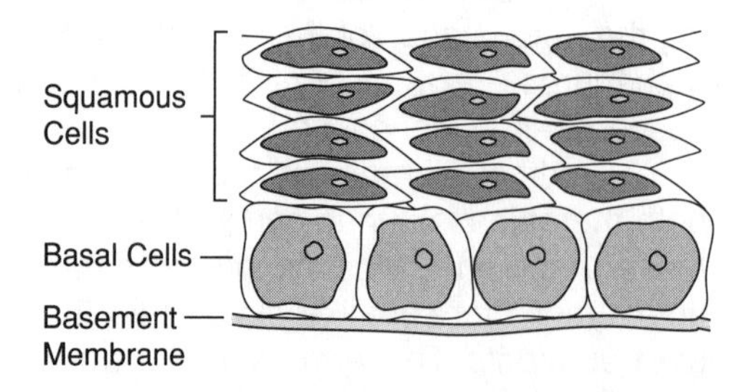

During a human papilloma virus infection, the virus makes its way through cracks in the skin or tears in the mucosal barrier until it reaches its target—a basal stem cell. Why the virus chooses this type of cell to infect is not completely understood, but one reason is that basal stem cells are proliferating. Indeed, viruses with limited genetic information typically target proliferating cells, because the cellular machinery required to make copies of the virus is already up and running in these cells.

As the infected basal stem cell proliferates, the virus within it is passed to its daughter cells. Some of these daughters become stem cells themselves, so by infecting basal stem cells which are self-renewing, HPV is able to establish a long-term infection that can last for months or even years. Other infected daughter cells take the elevator ride up to the surface, and by the time they reach the "top floor," many new viruses will have been produced and released from these cells. Importantly, human papilloma virus infections are "gentle" and do not kill the infected cells, so most HPV infections are unapparent.

Viral Spread

During an HPV infection, a continuous supply of newly made viruses is delivered to the surface of the skin or the mucosal barrier, allowing transmission to other humans. Human papilloma virus is spread either by direct physical contact with an infected individual or by direct contact with a surface on which the virus has been deposited. Transmission is facilitated by the fact that HPV's protein coat is impervious to agents which would destroy many other viruses: detergents, acids, heat, and drying. In fact, the detergents used in condoms, which can damage or destroy HIV-1 and herpes simplex viruses, are helpless against HPV. In contrast to herpes simplex infections, which are generally lifelong, most genital HPV infections are transient, persisting only for a few months or a few years.

Warts are the most visible manifestation of a genital HPV infection. However, the formation of a genital wart is just the tip of the iceberg, because most genital HPV infections are unapparent. Even a microscopic examination of cells taken from the vagina frequently will not detect the infection. For example, when a large number of college-age women were tested, almost half were found to be infected with genital HPV. Nevertheless, tissues which contained abnormal cells were detected in only about 3% of those who were HPV positive, and only about 2% of the infected women had genital warts. So although human papilloma virus infections of the genitals are common, genital warts are a relatively unusual outcome of such an infection.

Because most genital HPV infections are asymptomatic and transient, it has been difficult to determine how efficient sexual transmission actually is. However, because over a million cases of genital HPV infections are diagnosed each year in the United States, it is clear that certain types of HPV (including HPV-6, -11, -16, and -18) are spread very efficiently by sexual intercourse. In the United States, the incidence of genital HPV infections has increased dramatically over the last three decades, rising at a rate faster than that of genital herpes infections. In fact, human papilloma virus is now second only to chlamydia as the most commonly acquired sexually transmitted disease, with at least twenty-five million Americans currently infected.

Human Papilloma Virus and Cervical Cancer

In rare cases, infection with certain "oncogenic" types of human papilloma virus can result in cervical cancer (mainly cervical carcinoma). I say "rare," because less than 1% of women who are infected with genital HPV will ever suffer from cervical cancer. However, because so many women are now infected with this virus, HPV-associated cervical carcinoma has become the second most common cancer in women worldwide, causing about 250,000 deaths per year.

Although over a dozen types of HPV are classified as oncogenic, it is HPV-16 which is most consistently found in cervical cancers, with HPV-18, HPV-31, HPV-45, and others being found less frequently. Clearly, human papilloma virus infections alone do not "cause" cervical cancer, since the vast majority of genital HPV infections do not lead to cancer. Yet in over 90% of the cervical cancers that have been carefully examined, one or more of the oncogenic HPV types has been detected. So although an HPV infection is not sufficient to cause cervical cancer, in most cases it appears to be a necessary element in the development of this malignancy. The current hypothesis is that an HPV infection may increase the risk of developing cervical cancer, but other "insults" are required before an HPV-infected cell becomes cancerous. Although the mechanisms by which an HPV infection "facilitates" cervical carcinoma are not completely understood, the following is a likely scenario.

Two of the proteins specified by the HPV viral genes are called E6 and E7, and these proteins are required for a successful HPV infection. The viral E7 protein binds to the cellular tumor suppressor protein, pRB, interfering with its "restraint" function. As a consequence, the growth-factor pathway in which pRB participates is "always on"—just as if both copies of the cellular pRB gene had been mutated. Normally, proliferation of basal epithelial cells ceases soon after these cells detach from the basement membrane and begin to mature. However, when the viral E7 protein binds to pRB, infected basal epithelial cells can continue to proliferate during their ride up the elevator to the mucosal surface. Indeed, the continued proliferation of these cells is essential for a large number of new viruses to be produced.

Normally, unscheduled proliferation of maturing epithelial cells would be sensed by a safeguard system that includes the p53 tumor suppressor protein, and the errant cells would be triggered to commit suicide by apoptosis. However, because a dead cell won't produce much virus, the human papilloma virus has evolved a strategy to deal with the p53 safeguard system: The viral E6 protein promotes the destruction of the p53 tumor suppressor protein. The bottom line here is that to facilitate a viral infection, HPV effectively disables two of the cell's most important tumor suppressor proteins. As a consequence, a growth-promoting system involving pRB is activated inappropriately, and a safeguard system that depends on p53 is compromised. So the result of the viral infection is the same as if the cell had suffered disabling mutations in both copies of its pRB and p53 genes. Because loss of tumor suppressor function can contribute to a cell becoming a cancer cell, it is easy to understand why an HPV infection might be a risk factor for cancer.

It is still not completely clear why infection with some types of HPV (the oncogenic types) increases a woman's chances of developing cervical cancer, whereas infection with other HPV types does not. Analysis of the E6 and E7 proteins of oncogenic and non-oncogenic HPV types have shown that they <u>are</u> slightly different, and the current thinking is that the proteins of the oncogenic types are more effective at disabling the pRB and p53 tumor suppressor functions. But there is probably more to this story.

Although cells infected with HPV have suffered the equivalent of four "hits," having lost the function of tumor suppressor proteins specified by the two genes for pRB and the two genes for p53, this is not enough to turn a cervical epithelial cell into a cancer cell. Indeed, 99% of women who are infected with oncogenic human papilloma viruses never get cervical cancer. Additional cellular control systems must be corrupted before HPV-infected cells become cancerous—and this can take time. Fortunately, while most HPV-infected cells are waiting for those additional hits to occur, they are destroyed by the immune system, which is trying hard to wipe out the viral infection. That's why most HPV infections last only months, and that is also why most HPV-infected cells don't stick around long enough to accumulate the additional mutations required to produce cervical carcinoma. Consequently, it is only very rarely that an infected cell manages to evade the immune system long enough to allow the requisite control systems to be corrupted.

Cervical carcinomas usually arise decades after the initial HPV infection, and biologists aren't sure how HPV-infected cells evade the immune system for such a long time. One clue is that when cervical cancer cells are examined, the HPV DNA is found to have been inserted into one of the cell's chromosomes. How this cutting and pasting takes place is not understood, but during insertion, some viral genes usually are disrupted, and the proteins specified by these genes may normally serve to alert the immune system that the cell has been infected. Without these proteins to "look at," the immune system may view HPV-infected cells as uninfected, allowing them to survive immune surveillance.

It is important to note that although oncogenic HPV types can cause warts on the exterior genitals, most HPV-associated genital warts are actually caused by non-oncogenic human papilloma viruses (usually HPV-6 and HPV-11). However, because it is common for individuals to be infected with more than one type of HPV at a time, the presence of genital warts can signal a possible infection with oncogenic HPV.

Interestingly, genital herpes simplex infections were once believed to play a role in cervical cancer. However, because herpes DNA is not consistently found in cervical cancer cells, it is now thought that herpes simplex infections of the genitals, like external genital warts, are just "markers" for the types of sexual activities (e.g., having many sex partners) that can lead to infections with oncogenic HPV.

Screening for Cervical Cancer

As HPV-infected cells accumulate mutations, they usually progress through stages in which they become increasingly more cancer-like.

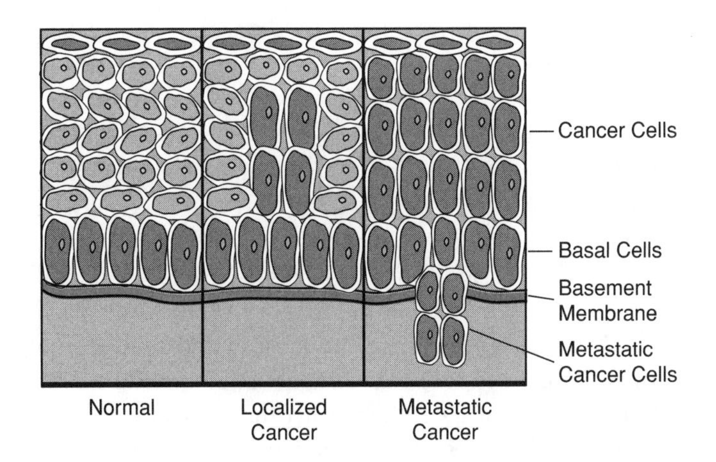

Although oncogenic human papilloma viruses can infect several other areas of the female reproductive tract (e.g., the vulva and the vagina), most HPV-associated cancers originate in the "transformation zone" of the cervix—the area where the reproductive tract changes from the multilayered epithelium of the vagina to the single-layered epithelium of the uterus. Why this region is a hot spot for cervical cancer is a complete mystery. However, because cervical cancer usually develops first in this one area, and because cervical carcinoma generally progresses in stages over a period of years, an examination of the transformation zone can reveal "weird" cells that are precursors of cervical cancer. A simple (and therefore inexpensive) way to examine these cells was introduced by a Greek immigrant, George Papanicolaou, for whom this procedure, the Pap smear, was named. This screen is so effective at detecting cervical cancer in the early stages that it is estimated that if all women were to have Pap smears performed every few years, the number of women who die of cervical cancer could be reduced by about 90%. Indeed, in countries where Pap smears are not available, about 70% of cervical cancer is not detected until it has reached an advanced stage—when it usually is impossible to cure.

The story behind the Pap smear and the man who invented it is quite interesting. Papanicolaou was born in Greece in 1883 and had received both an M.D. and a Ph.D. by the time he was twenty-seven. He heard stories about the opportunities available in the United States, and in 1913 he and his wife immigrated via Ellis Island with barely enough money to pay for their entry visas. This well-educated man began life in this country as a rug salesman, a clerk, and a violin player in a restaurant. Fortunately, he soon was hired as an assistant in anatomy at Cornell University, where he did research on the sex chromosomes of guinea pigs. To obtain cells for his work, he scraped cells from the reproductive tract of a guinea pig. He then smeared these cells on a glass slide, and examined them under a microscope. So it was on a guinea pig that the first Pap smear was performed! Perhaps this is where we get the expression "being a guinea pig."

A few years later, Papanicolaou decided to study the changes that take place in cells of the human reproductive tract over the course of a menstrual cycle. Again he used the "guinea pig smear" technique to obtain cells for his experiments. It so happened that one of the women he was studying was suffering from cervical cancer, and when he examined her smear, he could see that the cells he had collected were clearly abnormal. This gave him the idea that it might be possible to use the technique to screen for cervical cancer. He reported this observation at a conference in 1928, but nobody took it seriously. So about ten years later, Papanicolaou and a colleague decided to perform the procedure on every patient who came to the gynecology department of their hospital for treatment. In 1943, they published the findings from this study, which clearly showed that the smear could be used to diagnose cervical cancer in its early stages. Still, it was not until the early 1950s that Pap smears became widely used in this country.

If the Pap smear did not exist, roughly 40,000 American women would die each year of cervical cancer. However, because many women in this country do have regular Pap smears, this year only about 5,000 will die from this form of cancer. And if all women in the United States were screened regularly, it is estimated that this procedure would save the lives of about 36,000 women every year. I think we can agree that paying tribute to Papanicolaou by calling it the "Pap" smear is well justified!

Treatments for Cervical Cancer

The Pap smear identifies "weird-looking" cells that appear to be progressing toward cervical cancer. Depending on just how weird they look, a gynecologist may recommend a "wait and watch" strategy, with Pap smears being done more frequently. Alternatively, a biopsy may be performed to determine more precisely the potential of cells taken from the cervix to become cancerous, and based on these findings, the cancer can be staged. Patients may then elect to have the offending tissues removed by procedures which range from simply zapping the precancerous lesion with a laser to a radical hysterectomy in which the uterus, cervix, and part of the vagina are removed. If the cancer has not spread farther than the cervix and the upper part of the vagina, cure rates following surgery are around 85%. If the cancer has spread more extensively, but has not metastasized outside the pelvic region, it is usually treated with a combination of surgery, radiation therapy, and chemotherapy. Depending on the degree of spread, cure rates for cervical cancer at this stage range from about 10% to 70%. In its advanced stages, cervical cancer usually metastasizes to the bones, the lungs, or the bladder, and after metastasis to these distant sites occurs, the disease is generally fatal.

Cancer Prevention in the Future

Of course, the best way to deal with cancer is to prevent it from occurring. We have discussed the fact that a large fraction of human cancer could be prevented if people simply would refrain from smoking. Likewise, because essentially all cases of cervical cancer occur in women who have been infected with sexually transmitted types of the human papilloma virus, if people were content with having only one sexual partner during their lifetimes, cervical cancer could be eliminated. Unfortunately, although educating people about the health risks associated with smoking and having multiple sex partners can be useful, it is very difficult to get people to change their habits. So scientists are always on the lookout for ways to prevent cancer that don't involve lifestyle changes.

One way to prevent cancers that are associated with viral infections is through the use of a vaccine that prevents the infection. Vaccines are designed to give the immune system a "preview" of what a real infection will look like. That way, if an infection does occur, the immune system will be primed and ready to deal harshly with the infecting agent. Recently an experimental vaccine has been produced that is designed to protect against infection with one type of human papilloma virus, HPV-16. This particular type was chosen because it is one of the five HPV types that is most frequently associated with cervical cancer. In a study on a small number of healthy volunteers, this vaccine appeared to be safe, and produced, on average, levels of anti-viral antibodies that were about forty times as high as those detected during a natural infection with HPV-16. This initial success led to a larger trial which included about 1,500 women, half of whom received the real vaccine and half of whom received a fake (placebo) vaccine. When these two groups of women were followed for about a year and a half, none of the vaccinated group became infected with HPV-16. In contrast, forty-one of the placebo-vaccinated women became infected with this virus, and biopsies revealed that nine of these women had cells that showed early signs of becoming cancerous.

Although more testing is clearly needed, these results suggest that this vaccine may be effective in protecting against HPV-16 infections, and therefore may be useful in reducing the number of cases of cervical cancer. Indeed, if a vaccine could be prepared that would protect against the five types of HPV most frequently associated with cervical cancer, it is estimated that such a vaccination could reduce the incidence of cervical cancer about ten-fold and could, at least theoretically, prevent about 400,000 cases of cervical cancer annually worldwide. I say "theoretically," because most vaccinations are given by injection. This route of vaccination works well in countries that have well-developed health care systems, but in underdeveloped parts of the world, large-scale immunization via injection is problematic. To try to overcome this potential obstacle, experiments are now in progress to test whether humans can be vaccinated against HPV by inhalation—a vaccination route that would require neither sterile equipment nor extensive training of health care providers.

Liver Cancer

A second, excellent example of a cancer that has viral infection as a risk factor is liver cancer. The liver is a large organ (it weighs about three pounds!) located in the upper right-hand portion of the abdominal cavity on top of the stomach. One of the main functions of the liver is as a blood filter, and the liver gets the blood it filters from two sources. About 80% of the blood that flows through the liver comes from the portal vein, which collects nutrient-rich blood from the area around the intestines. The second source of blood for the liver is oxygen-rich blood, which is delivered directly from the heart through the hepatic artery. Indeed, about 25% of the output of the heart goes through the liver each time the heart beats. All this blood percolates through the liver, and then is collected and piped back to the heart to begin another circuit.

The plumbing within the liver is a little complicated, but basically the liver cells (hepatocytes) are positioned so that they have one "side" that is in direct contact with the blood that circulates through this massive organ. This allows a liver cell to take up toxins from the blood, and to detoxify them through various chemical reactions. The other "side" of each hepatocyte faces onto a bile duct, allowing the products of the detoxification process to be spit out of the hepatocyte, collected by the bile ducts, and poured into the small intestine for elimination with the feces.

This plumbing scheme puts the liver in a perfect location to detoxify chemicals in the blood, but it also ensures that any viruses which have entered the blood stream will pass right through the liver, and will come into intimate contact with the hepatocytes. Consequently, liver cells are prime targets for infection by many blood-borne viruses.

Two completely unrelated viruses have been identified which, when they infect liver cells, can dramatically increase the risk that infected individuals will eventually suffer from liver cancer. These viruses, hepatitis B and hepatitis C, get their names because they cause inflammation (that's the "itis" part) of the liver (the "hepa" part). About 70% of all liver cancer arises in one of the hepatocytes (there are other kinds of cells in the liver that also can become cancerous), and these cancers are classified as hepatocellular carcinomas. We will focus on hepatocellular carcinoma here because it is clear that viral infections are a major risk factor for this form of liver cancer.

Hepatitis B Virus

Hepatitis B virus is spread efficiently by blood-to-blood contact. In its most natural setting, this spread takes place on those occasions when the blood of an infected mother enters the blood stream of her baby during childbirth. Indeed, about 20% of babies born to hepatitis B-infected mothers will be infected at birth. I say "most natural setting" because, although hepatitis B virus can be efficiently spread, for example, when drug addicts share needles, the virus certainly did not evolve to be spread in this way. Another rather natural route of hepatitis B transmission is from child to child, probably through open sores or cuts. This is most likely to be a preferred route when many children are crowded together, and hygiene is lax (e.g., in some daycare centers).

Hepatitis B virus ranks as one of the most infectious of all viruses: Transfer of a fraction of a drop of blood is sufficient to spread the virus from one human to another. There are two reasons why blood-to-blood spread of hepatitis B virus is so efficient. First, the protective coat of hepatitis B virus is well suited to resist the assaults of molecules found in the blood which would destroy less-well-protected viruses. As a result, large quantities of hepatitis B virus can accumulate over time in the blood of an infected individual. In addition, the absolute favorite targets for a hepatitis B infection are the hepatocytes of the liver. The liver contains almost a trillion of these cells and is strategically positioned to intercept blood as it circulates through the body. Consequently, hepatitis B virus which has entered the blood stream is delivered right to its target.

Hepatitis B virus is also found in seminal fluid. Although sexual transmission of hepatitis B virus is common, it is relatively inefficient. Indeed, studies show that mates of infected individuals frequently are not infected even after years of continual sexual contact.

Chronic Infection with Hepatitis B Virus

Hepatitis B virus is one of the world's most important pathogens. Roughly two billion of the world's people have been infected with hepatitis B virus at some time during their lives, and about 500 million of these now carry the virus as a chronic infection. Over a million people die each year of hepatitis B-associated liver disease. In about 70% of infected adults, the immune response eradicates the virus, and the infection causes few or no symptoms. For the other 30%, the destruction of liver cells by the immune system is massive enough to cause the symptoms commonly associated with liver damage: nausea, vomiting, liver pain, jaundice, and dark-colored urine. These symptoms can last for several months while the immune system is battling to subdue the virus.

Fortunately, the immune system is usually victorious. In about 90% of symptomatic adults, the virus is eradicated, and damaged liver cells are replaced by the proliferation of healthy liver cells. However, in about 10% of symptomatic adults (roughly 3% of those initially infected), the hepatitis virus wins the battle with the immune system, and establishes a long-term, chronic infection.

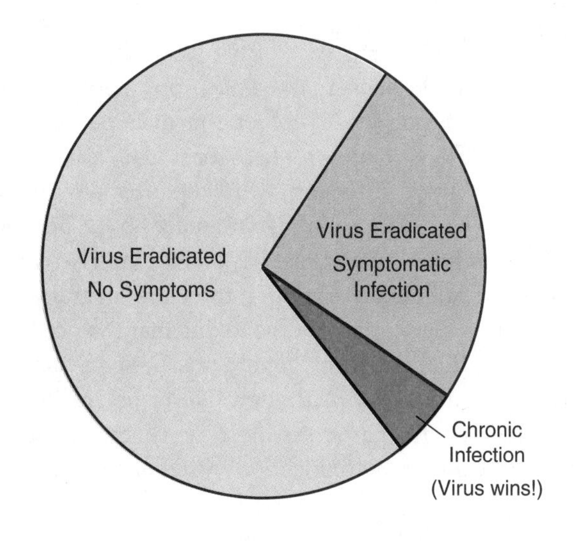

In newborn babies, these numbers are very different: About 90% of infected newborns end up as chronic carriers of the virus. This difference presumably reflects the immature state of the newborn's immune system. The high proportion of infants infected at birth who go on to become chronic carriers eventually provides a relatively large pool of mothers who can infect their offspring, passing the infection from generation to generation.

Across the globe, there are dramatic differences in the fraction of people who are chronically infected with hepatitis B virus. For example, in the United States, only about 0.5% of the population is chronically infected. In contrast, in parts of Asia (e.g., Vietnam), chronic infection with hepatitis B virus runs as high as 20%. There are two major reasons for these geographic differences. First, in the United States, donated blood is routinely screened for hepatitis B virus, so transfusion-related infections with this virus are very rare here. More importantly, however, vaccines against hepatitis B virus have been available in the United States since 1982, and the current vaccine is administered not only to health-care professionals, who routinely come into contact with blood and blood products, but also to children. In contrast, in those parts of Asia where chronic infection with hepatitis B virus is most common, donated blood is not routinely screened, and vaccines against hepatitis B virus are not readily available.

During a chronic infection, large amounts of hepatitis B virus circulate in the blood stream, and as uninfected liver cells proliferate to replace those killed by the immune response, these "fresh" liver cells can be infected by the circulating virus, perpetuating the chronic infection. Indeed, the impression one gets is that the immune system warriors march into the liver and try to deal with the virus-infected cells—but can't quite get the job done. Then, later, when the immune response has "relaxed," more cells become infected, and the cycle repeats.

Hepatitis B Virus-Associated Liver Cancer

The reason I make such an issue of chronic hepatitis B infections is that roughly 20% of long-term, hepatitis B carriers eventually develop hepatocellular carcinoma. Indeed, a chronic infection with hepatitis B virus increases one's risk of getting liver cancer about 200-fold. Exactly how this virus acts as a risk factor for liver cancer is not well understood, although there are several possible scenarios. First, many of the toxins that liver cells are designed to deal with are "genotoxins"—toxins that can damage cellular DNA. Although liver cells are stocked with enzymes that can detoxify these chemicals, sometimes the cellular detoxification systems can be overloaded. When this happens, hepatocytes become targets for the very genotoxins they normally protect against.

Unless your liver has been damaged, the cells of your liver don't proliferate. Consequently, hepatocytes usually have plenty of time to repair any damage inflicted by genotoxins. In contrast, during a hepatitis B infection, liver cells must proliferate to replace those killed by the immune response. This "extra proliferation" increases the risk that hepatocytes will copy their chromosomes and divide before DNA damage can be repaired. So the combination of being constantly exposed to genotoxins while being forced to proliferate may result in unrepaired DNA damage that can lead to cancer.

If exposure to toxins and increased proliferation were the whole story, however, you would predict that many of the liver cells which eventually become cancerous would be uninfected, "bystander" cells which proliferated to replace infected cells killed by the immune response. However, experiments have shown that for most hepatitis B-associated liver cancer, the cells that make up the tumor contain hepatitis B genes which have somehow been inserted into the chromosomes of these cells. This means that most hepatitis B-associated cancer arises in cells that have been infected with the virus, and suggests that the virus does something to these infected cells which actively contributes to the cancer-causing process. What might this be?

Most research designed to discover which hepatitis B viral functions might be involved in cancer has centered on the viral "X" protein, aptly named because its role in hepatitis B infections remains mysterious. Some experiments indicate that the X protein can bind to and alter the function of the p53 tumor suppressor protein. Although this finding is still controversial, if true, being infected with hepatitis B virus would be the equivalent of a cell sustaining mutations in both copies of its p53 genes. So by choosing to infect liver cells that constantly are subjected to DNA-damaging toxins, by causing increased liver cell proliferation, and perhaps by disabling the p53 tumor suppressor, hepatitis B virus may increase the probability that liver cells will collect the mutations required to transform them into cancer cells. Still, it usually takes many years for these mutations to accumulate, and hepatitis B-associated liver tumors generally arise twenty to fifty years after a person becomes chronically infected with the virus.

Hepatitis C Virus

Hepatitis C virus was first identified in 1989, but it isn't clear whether this virus is really a "new" virus, or whether advances in technology have only recently made its detection possible. Nevertheless, hepatitis C virus now infects about 170 million people worldwide, and in the United States, roughly 2% of the population is infected with this virus. This means that roughly five times as many Americans are infected with hepatitis C virus as with HIV-1.

Transmission of hepatitis C by "unnatural" routes involving the transfer of blood (e.g., by transfusions or by intravenous drug abuse) or blood products (e.g., clotting factors) is very efficient. This makes sense: Like hepatitis B, a large number of hepatitis C viruses are found in the blood of infected individuals, and hepatitis C virus infects the liver, through which large volumes of blood circulate continuously.

The "natural" routes of infection available to hepatitis C virus have proven more difficult to elucidate. Although the virus is found in saliva and semen, transmission of hepatitis C virus by sexual contact appears to be inefficient. Indeed, the only well-documented natural avenue of transmission is from infected mothers to their offspring when a mother's blood is introduced into the bloodstream of her newborn as a result of trauma during childbirth. However, only about 7% of children born to infected mothers are infected by this route, so there must be more to the story. In addition, roughly half of all hepatitis C patients claim not to have been exposed to blood or blood products. So it is likely that there are other routes of infection which have not been discovered—or that a lot of people have "forgotten" that they received a blood transfusion or that they injected drugs. So there is still a great deal of mystery surrounding the spread of hepatitis C virus.

Like hepatitis B, hepatitis C virus can establish a life-long, chronic infection, and over half of all individuals infected with hepatitis C virus become chronically infected "carriers." Indeed, in the United States, about 60% of newly diagnosed cases of chronic hepatitis are caused by hepatitis C infections. How hepatitis C virus manages to evade the immune system during a chronic infection is not well understood, but it is known that the machinery which hepatitis C uses to reproduce is very error-prone. Consequently, the virus mutates rapidly within infected cells, and therefore may be able to stay one step ahead of immune defenses.

The high rate at which hepatitis C virus mutates has another important implication: It is very unlikely that a vaccine can be devised which will protect against a hepatitis C infection. The purpose of a vaccination is to prepare the immune system for a potential attack by the "real thing." Hepatitis B virus rarely mutates, so the vaccine and the real thing are almost always the same. Consequently, an excellent vaccine has been prepared which protects quite effectively against hepatitis B infections. In contrast, hepatitis C virus mutates so rapidly that a very large number of slightly different hepatitis C viruses is circulating in the population—making it hard to imagine how a vaccine could protect against all these "variant" viruses.

Hepatitis C-Associated Liver Cancer

At least two-thirds of all new hepatitis C infections go unrecognized because they are asymptomatic or only mildly symptomatic. Nevertheless, about 2% of humans with chronic hepatitis C infections will eventually die from liver cancer (hepatocellular carcinoma), which usually arises about three decades post-infection. So hepatitis C often is billed as a "silent killer." Although the incidence of hepatitis B–associated hepatocellular carcinoma has been decreasing steadily over the past decade in this country, deaths from hepatitis C–associated liver cancer are increasing rather dramatically. Why a chronic hepatitis C infection predisposes a person to liver cancer is not known for certain, although there are reports that one of the hepatitis C proteins can associate with the tumor suppressor protein, p53, and interfere with its function.

Treatment for Liver Cancer

Cervical cancer progresses in stages over a number of years, and this progression can be monitored using Pap smears. When cancerous cells are detected, the cervix and associated tissues can be removed, if necessary, to eradicate the cancer. In contrast, although virus-associated liver cancer progresses slowly, and it is easy to detect hepatitis B or C viruses in the blood of infected individuals, chronic hepatitis C infections are difficult to cure, and it is currently impossible to cure chronic hepatitis B infections. Moreover, a functioning liver is required for life, so unlike cervical cancer, the target for infection by hepatitis B and C viruses cannot be removed. Surgery sometimes is used to try to eliminate the cancerous portion of the liver, but less than 50% of the patients who have this operation survive for more than three years.

Viral Infections and Cancer

In terms of their association with cancer, one of the most important features that human papilloma virus, hepatitis B virus, and hepatitis C virus have in common is their ability to establish long-term infections. Indeed, only viruses which cause long-term infections are associated with cancer in humans. This is because viruses only act as risk factors: They don't cause cancer all by themselves. It is believed that during long-term infections, cancer-associated viruses corrupt one or more control systems within the infected cell. This is clearly the case with human papilloma virus infections, and is likely to be true for infections with other viruses such as hepatitis B and C, which are associated with an increased cancer risk. The corruption of control systems in virus-infected cells probably reduces the number of mutations required to produce a full-blown cancer cell. However, oncogenic viruses must "stick around" in the cells they infect long enough for mutations which disrupt additional growth-promoting and safeguard systems to occur. For example, another virus which causes liver inflammation, hepatitis A, only causes short-term infections, and, as a result, infection with hepatitis A virus is not associated with liver cancer.

Thankfully, only a few viruses establish long-term infections in humans, and some of these (e.g., herpes simplex virus and the virus that causes chickenpox) are not associated with any cancer. So it is important to note that the ability to establish a long-term infection is necessary, but not sufficient for a virus to increase the risk of cancer. Indeed, there are only six viruses which, when they infect humans, are known to be risk factors for cancer: human papilloma virus, hepatitis B and C viruses, Epstein-Barr virus (Burkitt's lymphoma), human T-cell lymphotropic virus type I (leukemia), and HIV-1 (Kaposi sarcoma).

THOUGHT QUESTIONS

1. Do viral infections cause human cancer? Why or why not?
2. Infection with certain "oncogenic" viruses can be a risk factor for cancer. Give an example of how a viral infection can increase one's risk of getting cancer.
3. Why is it that the only viruses associated with cancer are ones which establish long-term infections?
4. Are all viruses which establish long-term infections risk factors for cancer? Why or why not?
5. What features of cervical cancer make it possible to screen for this malignancy? Why can screening reduce the number of deaths due to cervical carcinoma?
6. Why isn't screening for infection with hepatitis B or C virus used to try to decrease the number of deaths due to liver cancer?

Table of Concepts for Lecture 7

Concept	Example
Some viral infections are risk factors for cancer, but viral infections do not cause human cancer. Only viruses which maintain long-term infections can increase the likelihood of developing cancer.	Cervical and liver cancer
Multiple control systems must be corrupted to produce a cancer cell.	Cervical cancer
Some viral infections can corrupt cellular control systems, decreasing the number of mutations required for a cell to become cancerous.	Cervical and liver cancer
Some cancers progress in well-defined stages.	Cervical cancer
Screening procedures can detect certain cancers at an early stage.	Pap smear and cervical cancer

Cancer and the Immune System

REVIEW

Cancer of the cervix and liver cancer are two excellent examples of malignancies which have, as a risk factor, viral infections. A human papilloma virus (HPV) infection is associated with most, if not all, cases of cervical carcinoma, yet this virus cannot be said to "cause" cervical cancer. The reason is that the vast majority of women infected with HPV never get cervical cancer. So by itself, infection with this virus is not sufficient to cause cancer. The reason for this is now clear. Proteins specified by the human papilloma virus can disable the pRB and p53 tumor suppressors, resulting in the corruption of two important control systems in infected cells. Consequently, infection with this virus has the same effect as mutating both copies of the genes that specify these tumor suppressor proteins. Although this gives the infected cell a good "shove" in the direction of malignancy, two corrupted systems do not a cervical cancer cell make. More genetic damage must be done. Consequently, an HPV-infected cell must accumulate mutations in other genes that enable growth-promoting systems and disable safeguard systems before it becomes a full-blown cancer cell. Fortunately, most HPV-infected cells never accumulate these mutations, and as a result, less than 1% of women infected with human papilloma virus will develop cervical cancer.

The fact that HPV-infected cells must stick around long enough to accumulate mutations in control system genes explains why only long-term HPV infections are a risk factor for cancer. This is true of virus-associated cancer in general. If the immune system quickly deals with an attacking virus so that all virus-infected cells are destroyed, that virus infection will not increase one's risk for cancer. Only viruses like HPV and hepatitis B and C, which can evade immune destruction and establish long-term infections, are able to persist long enough to play a role in transforming an infected cell into a cancer cell. Thankfully, only a small number of viruses cause long-term infections in humans, and most of these do not produce proteins that can corrupt cellular control systems. Consequently, the number of viruses that are associated with human cancer is very small.

When routine Pap smears are used to screen for cervical carcinoma, this form of cancer usually can be detected early enough to be cured. Most cervical cancer originates in a limited region of the reproductive tract, so the area which must be sampled with a Pap smear is relatively small. In addition, although the cervix and surrounding tissues are important for reproduction, they can be removed, if necessary, to treat cervical cancer. In contrast to the favorable situation for early detection and treatment of cervical cancer, even when liver cancer is detected early, the entire organ cannot be removed: A functioning liver is required for life. This is an important consideration, because chronic hepatitis C infections are difficult to cure, and there is currently no cure for chronic hepatitis B infections.

CANCER AND THE IMMUNE SYSTEM

The most important concept to understand about cancer and the immune system is that the immune system did not evolve to deal with cancer. Our immune systems came into being to protect us against infectious diseases until we are old enough to reproduce and rear our offspring. Indeed, as recently as about 200 years ago (a blink of an eye on the evolutionary timescale), the life expectancy for humans was only about 35 years. Because cancer is a disease that usually afflicts older persons, most humans died of infectious diseases long before they were old enough for cancer to be a problem.

Internal Surveillamce

Fortunately, human cells come equipped with internal safeguard systems designed to protect against loss of growth control. Most of these safeguard systems evolved a very long time ago when single-cell organisms "banded together" to produce multicellular organisms. These safeguards were necessary to ensure that any cells which exhibited "antisocial" behavior would be dealt with harshly. Because cancer cells are notoriously antisocial, the same safeguard systems which evolved to help cells "get along" in multicellular organisms (like humans), also provide powerful and effective "internal" surveillance against wannabe cancer cells. Indeed, many of the proteins we now call tumor suppressors originally evolved as cellular "policemen," which would spring into action if one cell in a multi-cellular organism began to proliferate inappropriately.

External Surveillance by Immune System Cells

So every human cell comes equipped with safeguard systems that keep watch lest that cell lose control and become cancerous. The importance of these internal surveillance mechanisms is clear when we consider how a mutation in a single gene in a safeguard system (e.g., BRCA1) can greatly increase a person's chances of getting cancer.

In contrast, it is much less clear whether the immune system provides effective "external" surveillance against cancer, especially the types of cancer that are most common in humans. Indeed, because the "day job" of the immune system is to protect us against infection by parasites, bacteria, and viruses, any defense that the immune system provides against cancer is an unintended consequence of its ability to defend against infectious diseases.

There are, of course, many anecdotal (uncontrolled) reports of a connection between the "health" of the immune system and cancer. For example, we have all heard accounts of people who have been diagnosed with cancer at times when they were under great stress, and we suppose that stress somehow reduced the strength of their immune systems, allowing cancer cells to escape immune surveillance. We also have heard stories of patients with "incurable" cancer whose tumors vanished when they changed their diet or began to watch lots of cartoons. We imagine that their new diet or happy thoughts somehow strengthened their immune system, so that it was able to fight off the tumor. Unfortunately, the details of how the mind and the immune system might interact have not yet been worked out. Consequently, although there appears to be some connection between the human "psyche" and susceptibility to cancer, the role that the immune system might play in all this is not well understood.

In humans whose immune systems have been weakened (immunosuppressed), either by chemotherapy or by diseases such as AIDS, the increased incidence of lymphoma and virus-associated cancer is well documented. However, during immunosuppression, a similar increase in the most common human tumors—cancers that are not of blood cell origin and which are not virus-associated—has been difficult to demonstrate convincingly. Most studies have been done in patients who received organ transplants, and who were treated with immunosuppressive drugs to prevent their immune systems from rejecting the transplants. Some of these studies report unusually high numbers of certain cancers, but the type of cancer seems to depend in many cases on what organ was transplanted and what immunosuppressive drugs were used.

For example, one analysis of cancer in patients who received heart transplants revealed an incidence of lung cancer that was twenty-five times that of the general population. However, other studies (e.g., of patients who received kidneys) did not show such an increase in lung cancer. This leads one to ask whether lung cancer in heart transplant patients might be related to the fact that many of them were cigarette smokers (a known risk factor for heart disease). Similarly, another analysis of data from transplant patients showed a marked increase in the incidence of melanoma. However, this may be related to the use of the immunosuppressive drug cyclosporin, which is known to weaken cellular safeguard systems that repair UV-damaged DNA. Indeed, other transplant patients who did not receive cyclosporin did not have an increased incidence of melanoma. My interpretation of these data is that although the immune system may help defend against virus-associated and blood cell cancers, it probably is not a significant defense against most human tumors. Here's my reasoning.

Requirements for Effective Immune Surveillance

To be effective against an "enemy," the immune system must solve two problems. First, the system must somehow be alerted that an enemy is present. Immunologists say that the system must be "activated." Then, once the immune system has been activated, it must be able to identify the enemy. So to provide surveillance against cancer cells, the immune system must first be alerted that a cell has become cancerous, and then must be able to discriminate between a cancerous cell and a normal cell. For infectious diseases, the immune system has evolved clever ways of sensing danger and of identifying invaders which should be destroyed. In contrast, for most cancers, these two problems are very difficult for the immune system to solve.

Alerting the Immune System to Danger

Our immune systems include many weapons which, if they were used inappropriately, could lead to tissue destruction and disease. To help ensure that the powerful weapons of the immune system are used only against invaders, and that we don't "shoot ourselves in the foot," the "guns" of the immune system are kept "unloaded." Although our immune systems are always at the ready, "live ammunition" is not issued and the weapons are not deployed until unambiguous signals are received which indicate that there is danger.

Over millions of years, our immune systems have evolved to be quite adept at recognizing the danger posed by invaders such as bacteria, viruses, and parasites. This recognition is achieved through the use of receptors on the surface of immune system cells which are "tuned" to detect molecules that are characteristic of these invaders. For example, some immune system cells have receptors for a molecule called LPS which is found on the surface of many bacteria. This molecule is not normally present in the human body, so when it appears in the tissues during a bacterial infection, it is a clear indication that the body has been invaded by something foreign, and that defensive action is required. Likewise, virus-infected cells usually produce "warning molecules" which can bind to receptors on immune system cells, and can give a clear signal that a viral attack is underway.

Normally most cells of the immune system circulate through the blood and lymphatic systems, waiting to do their thing. These cells are "on call." When the alert is sounded, the guns are loaded, and the cells of the immune system are rushed to the scene of the infection. This response to danger signals is so efficient at mobilizing the immune system that most viral and bacterial attacks are dealt with even before we experience any symptoms.

Although the immune system responds very efficiently to danger signals that are characteristic of infecting microbes, the immune system has not evolved receptors designed to recognize tumor cells. As a consequence, immune system cells usually "don't get called," and just continue to circulate through the blood and lymphatic systems, oblivious of tumors that may be growing out in the tissues. So the first problem the immune system has in protecting us against cancer is that there is no efficient mechanism which can alert cells of the immune system to the danger that a tumor might pose.

Cancer Cell Recognition

The second problem the immune system has in providing protection against cancer is how to tell tumor cells from normal cells. The immune system is very good at recognizing cells that have been infected with bacteria or viruses, usually because the cells display "foreign" molecules on their surfaces. These foreign molecules identify infected cells as targets for immune destruction. In contrast, although some cancer cells do have unusual surface molecules which might distinguish them as cancerous, most cancer cells display only normal proteins on their surfaces (although sometimes in greater or lesser than normal amounts). Because cells of the immune system are "taught" not to react against our normal proteins (to protect against autoimmune disease), cancer cells with normal surface proteins (most cancer cells) are routinely ignored by the immune system.

Even for the rare cancer cells that do have distinguishing surface molecules, the immune system still has a problem in recognizing them as appropriate targets. Most cancer cells mutate rapidly, so within a growing tumor, there may be cells that have lost (due to mutation) the "clues" that the immune system might use to identify cancer cells. This problem is especially acute if the immune system "gets there late" (e.g., due to delays in being alerted to the danger) and has to try to deal with a large tumor which contains a huge number of mutant cells. Indeed, by mutating rapidly, cancer cells present the immune system with a "moving target," usually allowing them to stay one step ahead of immune destruction.

The Immune System and Blood Cell Cancer

Because immune system cells circulate through the blood and the lymph, where they may "rub shoulders" with cancerous blood cells, it seems possible that the immune system might provide some protection against blood cell cancers. Indeed, there is evidence that humans whose immune systems are suppressed do develop more non-Hodgkin's lymphomas than do humans with intact immune systems. However, this protection is clearly incomplete, since people with healthy immune systems still get blood cell cancers.

The Immune System and Virus-Associated Cancer

Immune system cells are very good at protecting us against viral attacks, so it might seem that the immune system would provide some protection against virus-associated cancers. However, there is a subtle point to be considered in evaluating the effectiveness of the immune system against cancers that have a viral infection as a risk factor. Only viruses which cause long-term infections are associated with cancer, and to establish a long-term infection, viruses must somehow evade destruction by the immune system. Consequently, you could argue that viral infections only become risk factors for cancer when the immune system has failed to provide protection against these infections. What this means is that although the immune system may be effective in decreasing the chances that a virus will establish a long-term infection (because the immune system may wipe out the virus), the immune system probably is ineffective against virus-associated cancers once they arise.

The Bottom Line

My conclusion is that, because the immune system was not designed to protect against cancer, it isn't very good as an anti-cancer defense. Although elements of the immune system may be of some help in dealing with cancerous blood cells and with viruses that are associated with some cancers, the more common "solid" tumors that arise in our tissues usually are beyond the reach of our immune systems. This is in striking contrast to the powerful internal surveillance provided by safeguard systems within each cell—systems which are essential for protecting us against cells that lose growth control and which might become cancerous.

Tricking the Immune System

Many tumor immunologists hold out hope that even though the immune system did not evolve to protect us against cancer, it might still be possible to "trick" the system into dealing with tumors. A large number of immune-based therapies are being tried, so I'll mention only a couple of them, just to give you an idea of what the possibilities might be. I think you'll see that although in some cases immunotherapies can be helpful, so far they are not as powerful as we might wish them to be.

Active immunotherapy

Two approaches are being used to try to trick the immune system into becoming a weapon against cancer: active immunotherapy and passive immunotherapy. The first is based on the idea that the immune system could play an active role in protecting against cancer cells if only it could be alerted that they exist. One strategy aimed at "waking up" the immune system is to vaccinate a patient against his tumor in hopes that the "prepared" immune system might then destroy the cancer.

A good example of this approach is provided by a recent Phase I/II trial involving 43 patients with pancreatic cancer. One gene that is frequently mutated in pancreatic cancer is the RAS proto-oncogene, so scientists produced the mutated version of the RAS protein in the laboratory, and injected it into these patients. The hope was that this vaccine might "get the attention" of the patient's immune system. Indeed, over half of the patients responded by producing immune system cells that could recognize the protein used for the vaccination. Most importantly, on average, those patients whose immune systems did respond to the vaccine survived more than twice as long as those who had no immune response to the mutant RAS protein. Although encouraging, these results must be interpreted with care. The number of patients in this trial was very small, and those who did not respond to the vaccine may simply have been "sicker" than those who did—and therefore likely to die sooner. In addition, the vaccination did not cure these patients. It only extended their lives by roughly three months.

Passive Immunotherapy

In the experiment I just described, the immune system of the patient was expected to play an active role in destroying his or her cancer. In a second approach, usually referred to as passive immunotherapy, scientists do most of the work. In one recent trial, scientists collected immune system cells (tumor-infiltrating lymphocytes) from melanomas that had been surgically removed from the skin of thirteen patients. They reasoned that these immune system cells were trying their best to destroy the melanoma, but there were just too few of these cancer-fighting cells to get the job done. So for each of these patients, the scientists grew their immune-system cells in the laboratory until they had about 100 billion of them, and then injected the cells back into the patient—and let them do their thing. The result was that the metastatic tumors of six of the patients did shrink, at least somewhat, and that for several of the patients, the shrinkage was dramatic.

Experiments like this suggest that, given assistance, immune system cells can be useful in treating some cancers. However, two points about this type of therapy should be mentioned. First, passive immunotherapy is extremely expensive, because the immune system cells must be isolated from each patient individually and then grown up in the lab—so this is a tailor-made treatment. This contrasts with standard chemotherapy, in which one chemical can be used to treat a very large number of cancer patients.

The second point to be noted is that most cancers display only normal proteins, so if the immune system is tricked into attacking cells that display this protein, there is a risk of producing autoimmune disease. For example, in the passive immunotherapy trial we just discussed, the protein (MART-1) recognized by the immune system cells is displayed on the surface of all normal melanocytes—but is more abundant on cells of many melanomas. And in this trial, the skin of most of the patients was "bleached," because the immune system cells destroyed the normal melanocytes as well as the cancer cells. Of course, white skin really isn't such a bad side effect (ask Michael Jackson), and this type of therapy might also be used for other cancers in which the target organ is not required for life (e.g., prostate, breast, thyroid, or ovaries).

Testing Immunotherapies

The results of these two trials of active and passive immunotherapy are fairly typical: There is frequently some positive response to immunotherapy, but it is almost always underwhelming. One reason for this "modest" response is that it is very difficult to test immunotherapies in the settings in which they are most likely to be of value.

For most cancers, a treatment already exists that has been shown to be of some benefit either in curing the cancer, or more usually, in extending the life of the patient. So to replace an existing treatment that is beneficial with an immune-based treatment that might not work at all would be unethical. As a result, new immunotherapies frequently are tested on humans under conditions that are far from ideal. For example, immune-based therapies are usually tested as a "last resort" on patients who have failed to respond to other cancer therapies. In these situations, the patients are generally very ill, and in a weakened condition as a result of earlier treatments. This problem is especially acute when testing active immunotherapies because, by their nature, these treatments depend on the patient's immune system being strong enough to fight the cancer.

Another way to deal with these ethical concerns is to use immunotherapy as an "add on" to an existing therapy, and to test whether the two therapies together are better than the existing therapy alone. This approach can also be problematic, because the existing therapy may render the immune-based therapy ineffective. For example, when immunotherapies are tested together with standard chemotherapy, the chemotherapy may weaken the patient's immune system to the point where the immune-based therapy has no chance of working. So giving cancer immunotherapy a fair test is a real problem. Usually, the new therapy must first show promise in one of these less-than-ideal situations before patients are willing to try the immune-based therapy as a primary treatment.

I think one of the settings in which we are most likely to see immunotherapies used in the future is in the treatment of "minimum residual disease"—what's left after the surgeon does his thing. This is in part because immune cells can attack cancer cells more effectively when there is not a huge tumor to deal with. Also, when there are relatively few cancer cells, it is less likely that the tumor will contain mutant cells which can resist the therapy. A specific immunotherapy might be effective, for example, following breast surgery to destroy the small number of cells that may have metastasized. Immunotherapy may also prove especially useful in treating slow-growing cancers, which tend to be less susceptible to radiation therapy and chemotherapy than are rapidly growing tumors.

THOUGHT QUESTIONS

1. The immune system did not evolve to deal with cancer. Why not?
2. Every human cell has powerful internal surveillance systems that protect against the cell becoming cancerous. Explain.
3. What two problems must the immune system solve to provide effective protection against cancer cells? Explain.
4. Give two reasons why the immune system is not very good at protecting us against the most common human cancers (solid tumors).

A CUMULATIVE TABLE OF CONCEPTS

The most important concepts we have discussed in Lectures 1-8 are listed in a cumulative table on the next page.

A Cumulative Table of Concepts

Concept	Example (Lecture #)
Cancer-associated mutations can result from translocations.	AML-ETO in acute myeloid leukemia (L2) BCR-ABL in chronic myeloid leukemia (L2)
Many cancers probably arise due to mutations that accumulate in immortal stem cells.	Acute myeloid leukemia (L2) Colon cancer (L6)
Inherited genes can predispose to cancer.	BRCA genes in breast cancer (L4)
Environmental factors can increase the risk of developing certain cancers.	Ionizing radiation and leukemia (L2) UV radiation and skin cancer (L5) Smoking and lung cancer (L5)
Some viral infections are risk factors for cancer, but viral infections do <u>not</u> cause human cancer. Only viruses which maintain long-term infections can increase the likelihood of developing cancer.	Cervical and liver cancer (L7)
Cancer results when <u>normal</u> control systems are corrupted. Cells don't need to learn anything new.	Breast cancer (L4)
Multiple control systems must be corrupted to produce a cancer cell.	Acute myeloid leukemia (L2) Colon cancer (L6) Cervical cancer (L7)
The same cancer can result from different mutations in different control system genes. Each cancer is a "family" of diseases.	Breast cancer (L4) Colon cancer (L6)
Control systems have multiple components, so there usually are several different genes that can be mutated to corrupt a system.	p16 in melanoma (L5) RB in lung cancer (L5)
Some viral infections can corrupt cellular control systems, decreasing the number of mutations required for a cell to become cancerous.	Cervical and liver cancer (L7)
Hormones can trigger cancer cell growth.	Breast and prostate cancer (L4)
Oncoproteins can short-circuit growth factor pathways.	BCR-ABL protein (L2) KRAS in colon cancer (L6)
Oncoproteins can block cell maturation.	AML-ETO protein (L2)
Oncoproteins can block normal cell death.	bcl-2 in follicular lymphoma (L3)

A Cumulative Table of Concepts (Continued)

Concept	Example (Lecture #)
Tumor suppressor proteins can help repair mutations.	BRCA proteins in breast cancer (L4)
Tumor suppressor proteins can perform "restraining" functions in growth factor pathways.	pRB in lung cancer (L5) p16 in melanoma (L5) APC protein in colon cancer (L6)
Tumor suppressor proteins can sense damaged DNA and halt proliferation until it is repaired.	p53 (L5)
Tumor suppressor proteins can oversee the orderly distribution of chromosomes.	APC protein in colon cancer (L6)
Some cancers progress in well-defined stages	Colon cancer (L6) Cervical cancer (L7)
Screening procedures can detect certain cancers at an early stage.	Mammograms and breast cancer (L4) PSA test and prostate cancer (L4) Skin cancer (L5) Colon cancer (L6) Pap smear and cervical cancer (L7)
Cancer cell development is clonal.	Breast cancer (L4)
Metastasis is a multi-step process.	Breast cancer (L4)
Different cancers usually have favorite sites to which they metastasize.	Breast and prostate cancer (L4) Lung cancer (L5) Colon cancer (L6)
Treatments can be tailored to mutation profile.	Chronic myeloid leukemia (L2) Acute lymphocytic leukemia (L2)
Drugs can heal corrupted control systems.	Gleevec for chronic myeloid leukemia (L2)
Bone marrow or stem cell transplants can be used to treat cancer.	Leukemia (L2) Follicular lymphoma (L3)
Monoclonal antibodies can be used to treat cancer.	Rituxan for follicular lymphoma (L3) Herceptin for breast cancer (L4)
Anti-hormone drugs can treat hormone-dependent tumors.	Breast and prostate cancer (L4)

A Cumulative Table of Concepts (Continued)

Concept	Example (Lecture #)
Cancers frequently contain mutant cells that can resist anti-cancer treatments.	Monoclonal antibody-resistant lymphomas (L3) Colon cancer (L6)
The immune system did not evolve to deal with cancer cells.	(L8)
The immune system may provide some protection against blood cell cancers and virus-associated cancers, but its role in protecting us from other types of cancer is limited.	(L8)

Cancer in the Future

REVIEW

Human cells have internal safeguard systems which provide extremely powerful surveillance against cells that lose growth control and become cancerous. In contrast, the external surveillance provided by the immune system is rather limited. The reason for this difference is that internal safeguard systems evolved to protect multicellular organisms against cells which exhibit antisocial behavior. Consequently, these same safeguard systems are great at protecting against cancer cells—cells which clearly are very antisocial. In contrast, the immune system evolved to protect us against invasions by infectious agents such as bacteria, viruses, and parasites. So if and when the immune system provides protection against cancer, it is an unintended consequence of the job it really evolved to do: protect us against infectious diseases.

The immune system's general ineffectiveness against cancer stems from its limited ability to solve two problems: how to know when cancer cells are present, and how to identify these cancer cells as being worthy of destruction. The immune system has evolved to recognize when we have been invaded by bacteria, viruses, and parasites. These infectious agents have "signatures" that the immune system has learned to recognize. In contrast, there has been no evolutionary pressure for the immune system to learn to recognize cancer cells, so they usually go unnoticed.

Even when the immune system is alerted that a tumor is growing, it has a real problem identifying cancer cells as targets. Most cancer cells look relatively normal on the outside, and the immune system is "trained" not to destroy our normal cells. Consequently, normal-looking cancer cells usually are ignored by the immune system. Some cancer cells do express unusual surface molecules, but even in this case the immune system generally is ineffective: Most cancer cells mutate so rapidly that a tumor contains a mixture of cells, some of which the immune system may recognize as being unusual—and therefore worthy of killing—and others which have mutated so that the "clues" the immune system might use to identify them as cancerous have been lost.

Because immune system cells circulate through the blood and the lymph where they are likely to encounter blood cell cancers, the immune system may provide better protection against cancerous blood cells than against solid tumors growing out in the tissues. Indeed, there is evidence that humans with compromised immune systems are more susceptible to blood cell cancers.

It has also been observed that immunocompromised humans have a higher incidence of some virus-associated cancers. This makes sense, because protecting against viral infections is one thing the immune system is very good at. However, only viruses which are able to establish long-term infections are associated with cancer, and to maintain such an infection, virus-infected cells must somehow evade immune detection. Consequently, the argument could be made that although the immune system may decrease the chances that a virus will get a "foothold" and establish a long-term infection, once virus-infected cells become cancerous, the immune system usually is powerless against them.

In summary, although the immune system may provide limited surveillance against some cancer cells, this is not what the system was designed to do. It is likely that the weapons of the immune system have some impact on the incidence of blood cell and virus-associated cancers. However, in the case of solid tumors that are not associated with a viral infection (the majority of human cancers), the protection provided by the immune system is usually a case of "too little, too late."

Although the immune system is not very effective as a defense against tumors, immunologists are hopeful that they can trick the immune system into performing "unnatural acts" which will lead to treatments for cancer. Two types of immunotherapy are currently being attempted. Active immunotherapy is based on the idea that if the immune system could somehow be alerted that a dangerous tumor was growing, it might be able to destroy the cancer. Passive immunotherapy builds on the premise that if immune system cells are removed from a patient and "enhanced" in the laboratory, these cells might become more effective anti-cancer weapons. Both types of immunotherapy are likely to be most useful in dealing with the small number of cancer cells that may remain after a surgeon has removed the primary tumor.

CANCER IN THE FUTURE

In this lecture, I will try to look into the future to see what strategies might be used to screen for and treat cancer a decade or so from now. Of course, unless you are Carnac the Magnificent, who knows the answers in advance, predicting the future is risky business—and I'm no Carnac. However this exercise will at least give us a feeling for what cancer biologists, immunologists, and oncologists are working on now. In addition, because the future always builds on the present, this will give us a chance to review some of the concepts we have already discussed. And review is good.

Screening for Cancer in the Future

If you can't prevent cancer from occurring, the next best thing is to detect it early, because cancer that is detected after it has metastasized to distant sites in the body is most often fatal. For example, here is a graph showing the percentage of patients who survived for more than five years when metastatic cancer was diagnosed in the years between 1975 and 1995. These survival rates are low, and you will notice that they have improved very little over this twenty-year period.

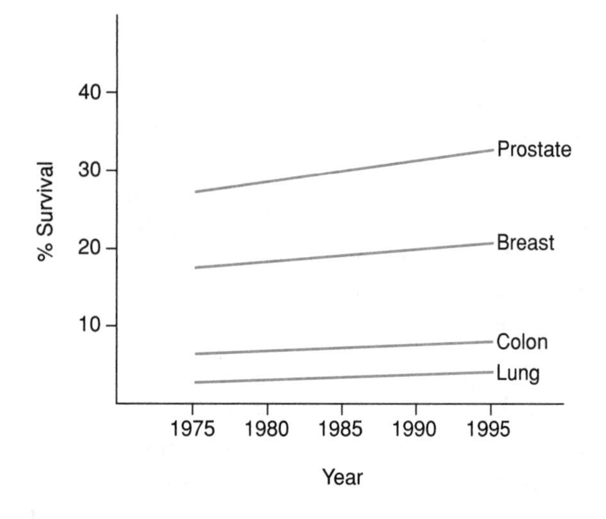

In contrast to cancer that has metastasized, cancer that is detected early usually can be cured. Consequently, in the future we can expect to see great emphasis placed on devising screening techniques which will allow the detection of incipient cancers.

Improved Cancer Imaging

Lung cancer is one of the malignancies that usually is not detected until after it has metastasized and is incurable, so a screen for early stage lung cancer might save many lives. One new imaging technique which may prove useful in screening for lung cancer is spiral computerized tomography. In this procedure, an x-ray source spirals rapidly around the patient, and a bank of detectors measures the fraction of these x-rays that penetrates the lung in any given direction. This data can then be used to construct a three-dimensional image of the lung. Such a scan takes less than thirty seconds and requires only a moderate exposure to x-rays. In its most advanced form, spiral computerized tomography can reveal tumors that are only about a quarter of an inch in diameter, so it is likely that this procedure could be used to detect lung tumors at a very early stage. However, there are several problems that must be overcome.

First, less than 10% of lung tumors smaller than half an inch in diameter ever will progress to become metastatic lung cancer. In addition, the lungs of some nonsmokers and the lungs of many smokers have nodules in this size range that are the result of damage due to a previous infection (e.g., a viral infection) and which have nothing to do with cancer. So there is a great risk of overdiagnosis when using spiral computerized tomography. One possible way to minimize overdiagnosis might be to compare images obtained several months apart to determine whether the suspicious "lump" is growing or changing in shape.

Another potential problem with early detection of lung tumors is that operations to remove these small cancers are extremely difficult to perform. Currently, most lung tumors that are removed surgically are larger than an inch in diameter. So to take advantage of early detection using spiral computerized tomography, a corresponding advance in surgical technique would be required. The National Cancer Institute is conducting an eight-year, $200 million study to determine whether early detection of lung tumors using spiral computerized tomography actually will decrease the number of people who die of lung cancer. This program should also gather information that may be helpful in discriminating between lesions that are dangerous and those that are not.

Protein Profiling

Another approach to early cancer detection is to examine readily obtainable bodily fluids (e.g., saliva, urine, or blood) for "telltale" proteins or other molecules which might signal that a cancer is starting to grow. We have already discussed a simple screen that measures the level of prostate specific antigen (PSA) in the blood. Using the PSA test, prostate cancer can be detected at an early stage when most of these cancers can be cured. Although more screens using single indicators like PSA will undoubtedly be developed, a lot of attention is being given to what might be called "protein profiling" in which a relatively large number of proteins are examined simultaneously to determine a "signature" that might forecast cancer.

In one recent study, proteins in blood samples collected from 50 women with ovarian cancer and 50 normal women were compared using a machine called a mass spectrometer. This sophisticated machine separates proteins according to their weight, and analyzes how much protein of a given weight is present in a blood sample. The protein signatures from these 100 blood samples were then compared using software that was originally developed by the military for identifying faint signals buried in a background of noise. The idea was to find patterns of proteins with various weights that were different in normal women vs. women with ovarian cancer. Indeed, the result of this "software training exercise" was the discovery of a protein signature that classified all fifty normal women as being cancer-free, and which identified all fifty of the women with ovarian cancer. The mass spectrometer and the trained software were then used to analyze the blood from 116 women, fifty of whom had already been diagnosed as having ovarian cancer. The results were pretty amazing. All fifty of the women with cancer were identified, and sixty three of the sixty six cancer-free women were also identified correctly. So in this small study, the false-negative rate was zero, and the false-positive rate was about 5%.

Although this result is quite impressive for a "first try," we need to put the rate of false positives in perspective. Ovarian cancer is fairly rare, afflicting only about .02% of women. So if women in the general population were screened for ovarian cancer using a test with a 5% rate of false positives, about 250 women (5/.02) would be identified as having ovarian cancer for every one who really did. That of course wouldn't fly. So although this screen might be useful for women who are known to be at high risk for ovarian cancer, or as a "second opinion" when another test suggests ovarian cancer, more work would need to be done before it could be used as a stand-alone screen.

One interesting feature of this type of screening is that it is not necessary to know which proteins make up the "cancer signature"—or what their functions might be. The "stupid" computer just looks for overall differences in protein patterns. This is contrasted with, for example, the screen for prostate specific antigen, in which both the identity and the function of this protein are known in advance.

In the future, it is very likely that many new screens will be developed which either alone, or in concert, will be very effective in spotting cancer early. However, it is important to note that screening to detect cancer at an early stage saves lives only if treatments exist which are more effective at curing cancer when it is detected early than when it has reached a more advanced stage.

Cancer Treatments in the Future

Scientists are working diligently to find new ways to treat cancer. Indeed, there are so many new approaches being explored that there is no way we could cover them all here. Rather, what I'd like to do in this lecture is to discuss a few of the strategies that are being tried, so you will at least get a feel for what is going on. Most cancers progress in stages, so we'll organize our discussion by asking how new treatments might halt the growth of cancers at various stages in their evolution. This will also give us a chance to review some of the important concepts we encountered in earlier lectures.

Turning Off Systems That Promote Growth

An early step in becoming a cancer cell is the corruption of systems that control growth, resulting in inappropriate proliferation. Because malfunctioning growth factor pathways are so common in cancer cells, drug companies are putting a lot of effort (and money!) into discovering drugs that can repair defective pathways. We saw an example of such a drug during our discussion of chronic myeloid leukemia. The leukemic cells of essentially all patients with chronic myeloid leukemia have a chromosomal translocation (the Philadelphia chromosome) that yields a BCR-ABL fusion gene. The protein specified by the ABL gene is a kinase enzyme that sticks a phosphate group on the next runner in its growth factor pathway. Normally, the ABL kinase is only active (i.e., only attaches a phosphate group) when a growth factor turns on the pathway. However, when the ABL gene is fused to the BCR gene, the ABL kinase is always active, resulting in leukemic cells that don't require growth factors to proliferate. Gleevec is a drug that has been developed specifically to inhibit the BCR-ABL kinase, thereby restoring these leukemic cells to health. However, because Gleevec is so specific, it is useful in the treatment of only a small fraction of human cancers.

In contrast to the BCR-ABL mutation, which is fairly rare, about 20% of all human cancers have mutations in a gene that specifies the KRAS protein. For example, about half of all colon cancers have KRAS mutations which result in the KRAS growth factor pathway being turned on all the time. In principle, if a drug could be developed which would reverse the effects of the KRAS mutation, it should be useful in treating a large fraction of human cancers. With this goal in mind, drug companies are trying to discover how to "fix" a defective KRAS gene. Unfortunately, these efforts have yet to yield a drug that is proven to be effective in treating cancer, although several candidate drugs are "in the pipeline."

Another possible way to "cure" cells with a KRAS mutation would be to inactivate one of the later runners in the relay. This runner would then be unable to pass the baton, even in cells where the mutant KRAS protein was signaling that the race was on. Two drugs, CI-1040 and BAY 43-9006, have been discovered that inhibit the action of two different proteins (MEK and RAF) in the KRAS relay. In Phase I trials, both drugs have shown some usefulness in treating several cancers. However, this approach has a potential limitation. Both the MEK and the RAF proteins are normal cellular enzymes that have important signaling functions in normal cells. Consequently, drugs which keep these proteins from functioning in cancer cells will also inhibit their function in normal cells. The hope, of course, is that cancer cells will be more sensitive to the loss of MEK or RAF function than are normal cells, but this remains to be seen.

Inhibiting Angiogenesis

Although an initial step in becoming a cancer cell is loss of growth control, for a solid tumor to become large enough to cause a problem, new blood vessels must be recruited into the growing tumor mass—a process known as angiogenesis. This can occur when systems within the cancer cell are turned on inappropriately, causing the production of angiogenic factors, which are released into the tissues surrounding the cell. Angiogenesis control systems and the factors they produce are attractive targets for anti-cancer drugs for several reasons. First, if angiogenic factors could be intercepted or inactivated, the tumor should stop growing, because it would be unable to acquire the blood supply required for continued expansion. Likewise, if new blood vessel growth could be stopped, even when angiogenic factors are present, tumors might stop growing or they might grow more slowly.

Angiogenesis inhibitors have another great advantage over many other anti-cancer therapies. Cancer cells make difficult targets, because they usually mutate so rapidly that the collection of cells in a tumor will likely include mutant cells that can resist most therapies. In contrast, new blood vessels are made up of normal cells that are not mutating rapidly. Consequently, drugs that either inactivate angiogenic factors, or inhibit their effects on blood vessel cells are much less likely to be rendered useless due to mutations.

Because drugs which inhibit new blood vessel formation might be used to treat a broad range of tumors, drug companies are trying hard to discover effective angiogenesis inhibitors. One angiogenesis factor that appears to be a promising target for therapy is a protein called VEGF. This protein is given off by many cancer cells, binds to receptors on the surface of the endothelial cells that line blood vessels, and encourages these endothelial cells to proliferate to make new vessels. Several monoclonal antibodies have now been produced which can bind to VEGF and prevent this protein from binding to its receptor. Initial results with one of these monoclonal antibodies, Avastin, have been encouraging. For example, in a Phase II trial of 116 patients with metastatic kidney cancer who had failed to respond to other therapies, Avastin was useful in slowing the growth of some of their tumors. Trials are now in progress to test whether more dramatic effects can be achieved by combining Avastin with other anti-cancer drugs.

In an adult human, very little angiogenesis takes place. Our vascular systems are quite stable, and although the growth of new blood vessels sometimes is required to replace those that have been damaged (e.g., due to a wound), adults could probably get along with little or no angiogenesis—at least temporarily. It is this thinking that has prompted cancer biologists to try to identify markers that could differentiate between old and new blood vessels. If this could be accomplished, it might be possible to selectively kill off new vessels while sparing the old ones. Again, this approach has the advantage that the target of the therapy would not be the cancer cell itself, which could mutate to resist the treatment, but normal blood vessel cells that mutate much less frequently. Although no drugs that selectively target new blood vessels are being tried yet on humans, recent experiments on mice suggest that this approach might be a good one. Here's how it works.

In the mid-1990s biologists discovered that a protein called an integrin was expressed on the surface of cells that make up new blood vessels, but not on the surface of cells in most established blood vessels. This same integrin protein is used by several different viruses to gain entry into cells. Indeed, when these viruses bind to the integrin proteins on the cell surface, they are transported right into the cell! For these reasons, it seemed that the integrin protein might be just the ticket to target new blood vessel cells, and to deliver some sort of poison into these cells. But what kind of poison should it be?

When angiogenic factors like VEGF bind to receptors on the surface of blood vessel cells, these cells are triggered to proliferate to form new blood vessels. And one of the important "runners" in the growth factor relay that conveys this "proliferate" signal is the RAF protein I just mentioned. Biologists reasoned that if they could somehow disrupt the function of RAF proteins inside the cells that make up new blood vessels, these cells would be unable to proliferate in response to angiogenic factors like VEGF. To make this happen, the biologists designed a gene that directs the production of a defective RAF protein which can "receive the baton," but can't "pass the baton" to the next runner—thereby stopping the growth factor relay in its tracks. If this "poison" RAF protein could somehow be introduced into growing blood vessel cells, these cells should cease proliferating, even when a tumor was giving off angiogenic factors. And because the RAF protein also functions as a survival factor for blood vessel cells, disrupting RAF function might even cause the targeted cells to die.

To test this strategy, scientists coated synthetic particles with integrin proteins (to target new blood vessels) and loaded these particles with many copies of the poison RAF gene (designed to shut off the VEFG growth factor pathway). When these particles were injected into mice that had large tumors, sure enough, the blood vessels supporting the tumors began to die. Moreover, within six days, some of the tumors vanished and others were drastically reduced in size. Here, the word "large" is important, because if scaled to the size of a human, these mouse tumors would weigh roughly four pounds! So this wasn't just a few cancer cells being killed.

Of course, we are not talking about curing human cancer here. These are cancers growing in mice. However, the experiments with the mutant RAF gene and the integrin protein illustrate an important point: The people who are doing these experiments are very clever, and they are trying every approach they can think of to discover new cancer treatments.

Inhibiting Early Steps in Metastasis

For most cancers to become really dangerous, they must metastasize to distant parts of the body, and form secondary tumors that are too numerous (or too difficult) for surgeons to remove. Consequently, if metastasis could be inhibited, many cancer patients could be saved.

For cancer cells to metastasize, control systems must be corrupted so that these cells can become mobile. At the time when an embryo is forming, cells move around a lot to get themselves into position to produce the requisite organs. So the control systems that are required for mobility are normal cellular systems. However, once the embryo has been formed, mobility systems generally are shut down, and most cells stay put for the life of the individual. Indeed, there are safeguard systems that enforce the sedentary nature of most normal cells. For example, the cells that make up our tissues are embedded in a mesh made of proteins and carbohydrates called the extracellular matrix. Cells attach themselves to the extracellular matrix using proteins on their surface that bind to various components of the matrix. To keep cells from wandering off on their own, normal cells in the tissues are programmed to commit suicide by apoptosis if their connections with the matrix are severed. So the binding of cells to the extracellular matrix not only gives them physical support, but also provides survival signals which keep them alive. What this means is that for a cell to metastasize, the safeguard systems that enforce contact with the extracellular matrix must be disrupted, allowing the cell to travel without dying.

In general, cells move by extending part of their cell bodies in the direction they wish to travel. If they are able to grasp the extracellular matrix with this extended "hand," they can then retract the hand and pull the rest of the cell along. It looks something like this:

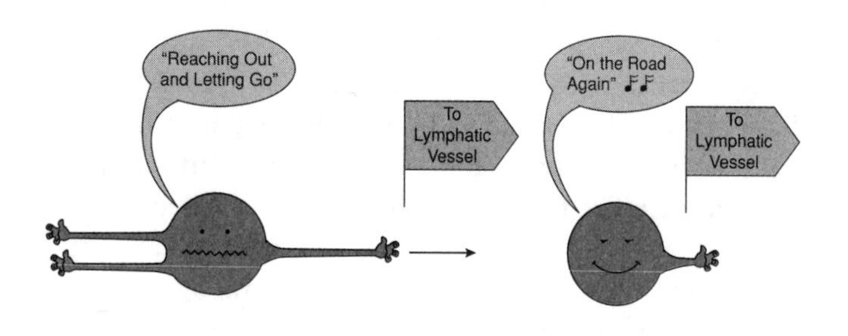

For this "scoot-along" trick to work, however, two things must happen. First the "hand" must be able to get a firm grip on the extracellular matrix, and second, the cell must let go of the other contacts that are anchoring it in its original position. Scientists have observed that in many cancer cells, integrin molecules are produced and positioned on the cell surface in the region that is destined to become the leading edge of the cell—the "hand," if you will. In essence, the integrin proteins form the "fingers" that can make strong attachments to the extracellular matrix. This concentration of integrin proteins at the leading edge of the cell gives the cancer cell the firm grip it needs to pull itself forward. Moreover, the binding of integrin molecules to the extracellular matrix initiates a survival signal that keeps the traveling cell from committing suicide when contacts with the extracellular matrix are severed to allow it to leave its old position.

So the inappropriate expression of integrin proteins on the surface of cancer cells does two things: It allows the cancer cell to pull itself along, and it provides survival signals to "fool" the safeguard system that would destroy a normal cell which tried this maneuver. With this in mind, scientists reasoned that integrin proteins might make a stellar target for drugs that could inhibit this important step in metastasis. Indeed, two drugs (Vitaxin and Cilengitide) which can bind to integrin proteins and prevent them from forming an attachment to the extracellular matrix are now in Phase II trials. Drugs of this type may prove useful in preventing early stage cancers from metastasizing.

Inhibiting Later Steps in Metastasis

Once control systems have been corrupted, allowing cancer cells to move from their normal positions in the body, wannabe metastatic cells are faced with the next problem: how to gain entry to the blood or lymphatic systems which can transport the cancer cells to distant sites. One of the enzymes given off by many cancer cells—a matrix metalloproteinase—is believed to play an important role in accomplishing this feat. A proteinase is an enzyme that cuts up certain proteins, and a metalloproteinase is a proteinase which requires the assistance of metals (e.g., zinc) to do this cutting. Interestingly, inappropriately expressed metalloproteinases are responsible for the destruction of the elastic fibers in lungs of cigarette smokers, leading to emphysema.

Metalloproteinases are skilled at cutting up proteins found in the extracellular matrix, and it is thought that the inappropriate expression of these enzymes helps cancer cells cut their way into the blood or lymphatic system. A number of new drugs have been discovered which can inhibit matrix metalloproteinase function. For example, a drug called Marimastat showed some benefit in extending the lives of patients with advanced cancer in a recent trial. Unfortunately, for ethical reasons, the clinical trials to test these matrix metalloproteinase inhibitors have been done on patients with advanced metastatic disease (i.e., well after considerable metastasis had already occurred). In contrast, these new drugs are most likely to be effective in preventing metastases when cancer is first diagnosed. Consequently, we might expect that in the future, trials will be conducted to test whether anti-metastasis drugs can reduce the likelihood that early stage cancers will metastasize.

The current thinking about metastasis is that getting into the lymphatic or blood systems and being carried to distant sites really is not the hardest part for wannabe metastatic cancer cells. The major obstacle for these cells seems to be coping with the environment of their adopted homes. Indeed, it is estimated that only a few percent of traveling cancer cells is able to grow when they reach distant sites, and of these, only about 1% acquires the new network of blood vessels needed to form a metastatic tumor. Because wannabe metastatic cells have such a hard time "settling in," scientists reason that by making their new homes even more inhospitable, these traveling cancer cells might be prevented from ever forming a tumor.

One way to do this might be to use angiogenesis inhibitors to prevent metastatic cells from forming tumors of significant size. For example, after a woman's breast cancer has been surgically removed, it might be useful to treat her not only with radiation to "mop up" any cancer cells that remain at the site of the primary tumor in the breast, but also with angiogenesis inhibitors. One potential problem with this approach is that cells which metastasize to distant sites can lie dormant or can grow slowly, so the need for new blood vessels to support the growth of a tumor may arise as long as a decade after the initial "seeding" of metastatic cells. Consequently, angiogenesis inhibitors might have to be taken for a long time, and this may not be compatible with good health. Because drugs that inhibit angiogenesis are so new, the usefulness of these drugs in preventing the formation of metastatic tumors remains to be determined.

Advanced Combination Treatments

In the future, I think we will see more use of combination treatments in which several different chemical and/or immune-based therapies are employed simultaneously. Clearly, a cancer cell's best defense against almost any treatment is its ability to mutate rapidly. Most of the time, treatment of a tumor with a single therapeutic agent ultimately proves ineffective, because a tumor is composed of cells with various combinations of mutations—and somewhere within that tumor, there is a cell or cells which have the mutations required to resist that particular agent. Consequently, although the tumor may decrease in size (regress) as sensitive cells are killed, resistant cells will continue to proliferate and restock the tumor.

Combinations of several different standard chemotherapeutic drugs are currently being employed with some success to treat certain cancers. However, if several radically different types of therapy (e.g., an angiogenesis inhibitor, a chemotherapeutic drug, and a specific immunotherapy) were used simultaneously, it is much more likely that no cells within a tumor would be able to resist the effects of the combined treatments. As more different types of weapons become available, this sort of "advanced combination therapy" will probably become the norm.

Personalized Medicine

In the future, I believe that cancer will become a much more "personal" disease. This personalization of cancer will take many forms, including individual risk analysis, tailor-made treatments, and patient-specific outcome assessment—all based on the genes present in each person's cells.

Genetic Profiling to Predict Cancer Risks

In Lecture 4, we noted that mutations in two genes, BRCA1 and BRCA2, can predispose a woman to breast cancer. If a woman is in a high risk family, she may choose to have her genetic profile done to determine whether the DNA of her cells contains a mutated BRCA gene. As more susceptibility genes are discovered, it will be possible to use genetic profiling to alert a person to "be on the lookout" for certain cancers. For example, if a person knew in advance that, because of his genetic makeup, he was particularly susceptible to exposure to certain kinds of carcinogens, he could attempt to limit that exposure. This would be functionally similar to the profiling that is now used to identify risk factors (e.g., obesity, stress, high blood pressure, and family history) that might predispose an individual to heart disease, allowing that person to modify his diet or lifestyle in order to avoid a heart attack.

Equally useful, an analysis of a person's susceptibility genes will give information about certain cancers which a person is very unlikely to get. For example, if a woman's genetic profile indicates that the chances of her getting breast or colon cancer are very small, she might choose to forgo mammograms and colonoscopic examinations. In the future, genetic screening to determine a person's susceptibility profile will likely be an important element of preventive medicine.

All Cancers Are Not Created Equal

Even when we focus on one particular cancer (e.g., breast cancer) we are really dealing with a collection of different diseases. There are two main reasons for this. First, every control system within a cell is composed of multiple components, just as the heating system in your home is a multi-component system. Consequently, there are usually a number of different ways that a given control system can be corrupted. For example, when we discussed melanoma, we noted that three proteins, pRB, p16, and CDK4, are components of the same growth-promoting system, and that a mutation in any one of these proteins could corrupt this system. However, these mutations may not be equivalent. The reason for this is that Mother Nature is quite frugal, and she frequently uses a single protein to perform multiple functions in a cell. So although pRB, p16, and CDK4 do work together in one growth factor pathway, each of these proteins also may play a role in other control systems. For example, mutating RB may disable system A (the system pRB, p16, and CDK4 have in common) plus system B, whereas a mutation in the gene for p16 may disable systems A and C. Consequently, the properties of a cancer cell with an RB mutation may be somewhat different from a cancer cell with a p16 mutation. Both mutations may result in melanoma, but the melanomas may have different characteristics.

So a given control system can be corrupted in different ways (due to different mutations), producing cancer cells that have different properties. But there is a second reason why all cancers of one type may not have the same characteristics: There are usually several different combinations of control systems which can be corrupted to make a cell cancerous. For example, when we discussed colon cancer, we noted that for a polyp to form, a wannabe colon cancer cell must suffer a mutation in a growth-promoting system, triggering inappropriate proliferation. In about half of all colon cancers, this control system is one in which the gene for the KRAS protein has been mutated. However, in other colon cancers, the control system that includes KRAS remains healthy, while another of the growth-control systems is corrupted. In both cases, corruption of control systems leads to colon cancer, but the properties of these colon cancer cells may be different, depending on the particular control systems which are corrupted.

Basing Treatments on Genetic Profiles

The fact that all breast cancers or all colon cancers or all melanomas are not alike has important implications for the treatment of cancer. Currently, most cancers are treated on a statistical basis. For example, for a Dukes stage B1 colon cancer, the treatment most likely to cure the "average" patient is surgery plus radiation therapy. However, there is really no such thing as an average cancer patient, because each cancer will have a unique set of mutations. And when all patients are lumped together, some may be overtreated, some may be undertreated, and some may not receive the appropriate treatment at all. Consequently, in the future, cancer treatments will become much more personalized, with treatments tailor-made for each patient.

Advances in molecular biology have already made it possible to compare the genes present in a patient's normal cells with the genes in his cancer cells (obtained, for example, by biopsy) to determine which mutations the cancer cells have suffered. Such studies can yield a genetic profile which can be helpful in designing treatments for individual patients. When we discussed childhood acute lymphocytic leukemia at the end of Lecture 2, I mentioned that one reason for the recent success in curing this form of leukemia is that treatments are now being tailored to fit the patient. Those children with AML1 mutations are treated in a different way than are children whose leukemic cells have the Philadelphia translocation. Indeed, in the future it is likely that the first question an oncologist will ask before he begins treating any cancer patient is, "What is the genetic profile of the cells in this patient's tumor?"

There are several different ways this information can be used to personalize cancer treatments. For example, genetic profiling can be used to identify the control systems and even the elements within these systems that are corrupted, making it at least theoretically possible to repair the damage. Genetic profiling is already being used to identify patients whose leukemic cells have the BCR-ABL mutation, because these individuals are candidates for treatment with Gleevec—a drug which can inhibit the deadly function of the BCR-ABL oncogene.

I think it is important to point out, however, that trying to cure cancer by "repairing" a single gene or a single pathway may be a difficult task. Oncologists have noted that many patients treated with Gleevec respond initially, but then become resistant to the beneficial effects of the drug. This problem arises because tumors contain a mixture of cells with many different mutations. Within this collection there are likely to be cancer cells with several different combinations of corrupted control systems. A drug like Gleevec may repair one control system, but there are likely to be cells in the tumor that don't require corruption of this particular system to become cancerous—they do it another way. And when the smoke clears from the attack of Gleevec on the susceptible cells, the resistant cells emerge.

Using Genetic Profiling to Predict Outcomes

Cancer treatments will also be personalized by using genetic profiling to predict whether or not an individual patient's tumor will metastasize. This is an important question. Many of the standard treatments for cancer can be devastating, and if these treatments are unlikely to be useful, it would certainly be nice to know that in advance. Indeed, one of the features of cancer that has puzzled oncologists is that patients with seemingly similar tumors often have very different outcomes. For example, some patients with prostate cancer will live quite happily for many years without treatment because their tumors will remain within the confines of the prostate. These individuals will eventually die from some other cause (e.g., "old age"). Other patients, with tumors that appear to be similar, will succumb rather quickly because their prostate cancer will metastasize to distant sites in the body. This has spawned the saying that although most men who live long enough will get prostate cancer, some will die from it, and others will die with it.

Recently there have been advances in genetic profiling that may help predict the outcome in patients with several types of cancer. For example, biologists screened about 25,000 human genes and identified seventy that were expressed at dramatically different levels in metastatic vs. nonmetastatic breast cancer. It was hoped that this profile of gene expression might be useful in predicting which breast cancers are more likely to metastasize. To test this idea, they examined breast cancers from 295 women, and concluded that 180 fit the "likely-to-metastasize" profile. Ten years after these samples were taken, the scientists recorded whether metastasis indeed had occurred for each of the women tested. About 50% of the women with the likely-to-metastasize profile did have detectable metastasis to distant sites. In contrast, only about 15% of the women who did not fit this profile developed metastases. Of course, in an ideal world of personalized medicine, you would like a genetic profile to predict with 100% accuracy whether a given patient's cancer would metastasize. However, in this trial, genetic profiling already gave a better indication of the likelihood of metastasis than did the standard predictors such as whether or not the primary breast tumor had spread to nearby lymph nodes.

In the future, cancer treatments are likely to become more and more personal, with genetic profiling being used as an important tool to determine who is at risk for certain cancers, which control systems are corrupted and should be repaired, and which treatments are most appropriate for each individual patient.

Carnac Predicts

As I'm sure you understand, cancer is a very tough problem. Some of our best minds have been trying for decades to find better ways to treat cancer—with relatively little success. The main reason this problem has been so difficult to solve is that most cancer cells mutate rapidly. This major obstacle notwithstanding, I believe there are several areas in which considerable progress will be made in the battle against cancer in the next decade.

Most deaths from cancer are preventable. Consequently, I foresee that education about avoidable risk factors (e.g., cigarette smoking, asbestos, and radon) and vaccines against virus-associated cancers will be effective in reducing the incidence of several of the most important cancers—especially in developed countries such as the United States.

If cancer is detected early, when the collection of mutant cells is relatively small, treatments are much more effective. So as I gaze into my crystal ball, I see that in the not-too-distant future, there will be "cancer screening centers" where a person can undergo a battery of tests designed to detect many different types of cancer at a stage at which they easily can be cured.

Most of the treatments we have discussed require an understanding of how cancer cells work—an understanding that is far from complete. In the future, basic research on growth-promoting and safeguard systems will suggest additional targets for anti-cancer drugs. Moreover, a more detailed understanding of how these growth-control systems interact with other cellular systems will be useful in predicting how anti-cancer drugs might affect normal cells, causing unwanted (and unexpected) side effects. These studies will help translate a theoretical understanding of cancer concepts into therapeutic advances.

Finally, I see a future in which simultaneous treatments with multiple therapeutic agents will be the rule rather than the exception. These combination therapies will be used to treat the minimal residual disease that remains after surgery, and some of these therapies will be tailor-made to fix corrupted control systems that are revealed by genetic profiling. The key here will be to attack simultaneously with patient-specific weapons at a time when there are relatively few cancer cells which need to be destroyed—and while the patient is still relatively healthy. Even under these "ideal" conditions, however, it's going to be a tough battle—but one that is certainly worth fighting.

Glossary

Angiogenesis – The process of growing new blood vessels.

Angiogenic factors – Substances (e.g., proteins) that stimulate the growth of new blood vessels.

Apoptosis – Sometimes called programmed cell death. The process by which cells commit suicide in response to problems within the cell or to signals from outside the cell.

Basement membrane – The collagen-rich membrane that underlies all epithelial surfaces.

Blast cell – An immature cell that proliferates rapidly.

Carcinogen – An agent that can play a role in transforming a normal cell into a cancer cell. Carcinogens usually work by causing mutations.

Carcinoma – Cancers that arise from one of the epithelial cells that line the surfaces that protect our bodies from the outside environment.

Carcinoma *in situ* – A carcinoma that has not yet broken through the basement membrane, and therefore still is located "at the site" where it originated.

Engraftment – The process by which donated stem cells find their way into the marrow of the recipient.

Epithelial cells – The cells that line our body cavities (e.g., colon epithelial cells) and the skin cells that cover our bodies.

Genotoxin – A toxin that causes damage to DNA.

Graft vs. host disease – A disease that occurs when killer T cells in donated bone marrow attack the cells of the recipient.

Graft vs. leukemia effect – A good result that occurs when killer T cells in donated bone marrow attack the leukemic cells of the recipient.

Growth factor – A protein that causes a cell to proliferate, usually by binding to a receptor on the surface of the cell.

Growth factor pathway – A series of proteins that form a "relay" to convey the "start proliferating" signal from the surface of the cell, where the growth factor binds —to the nucleus of the cell, where the genes that control growth are located.

Hepatocyte – A liver cell.

Hybridoma – A hybrid cell made by fusing a cancerous mouse cell with a mouse B lymphocyte (plasma cell) which is making antibodies. The hybridoma that results is an immortal antibody factory.

Lesion – A rather generic term for an injury.

Leukemia – A cancer which results when a blood cell in the bone marrow proliferates more rapidly than its normal counterparts and fails to mature properly.

Loss of heterozygosity – This term is usually used to describe what happens when both copies of a tumor suppressor gene suffer mutations such that the proteins they specify no longer function.

Lymphoma – A cancer that arises when a lymphocyte (either a B or a T lymphocyte) which has left the bone marrow fails to mature properly.

Major histocompatibility complex proteins – A group of proteins that are found on the surfaces of our cells and which are so different from human to human that they make transplantation difficult.

Malignancy – A synonym for cancer.

Mobilization – The process of treating a person with drugs which increase the number of stem cells in his blood.

Mutation – A change in the DNA that makes up our chromosomes.

Myeloma – A cancer that arises when a B lymphocyte fails to die when its useful life is over.

Oncogene – A mutated proto-oncogene that specifies a protein (an "oncoprotein") which, by its action, can contribute to a cell becoming a cancer cell.

Oncoprotein – A protein which, by its action, can contribute to a cell becoming a cancer cell.

Pathogen – A microbe (e.g, a bacterium or a virus) that causes disease.

Phosphorylation – The act of adding a phosphate group to another molecule.

Plasma cell – A B lymphocyte that produces antibodies.

Proliferation – The process in which cells grow to twice their original size, and then divide to produce two daughter cells.

Proto-oncogene – A gene which, if mutated, can become an oncogene.

Stem cell – A self-renewing cell that helps resupply the body. For example, blood stem cells resupply the blood system when blood cells die, and colon epithelial stem cells resupply the surface of the colon when cells there are destroyed by normal wear and tear.

Susceptibility gene – A gene which, if inherited, can make a person more or less susceptible to developing cancer.

Transcription factor – Proteins that can regulate the levels at which cellular genes are expressed (i.e., that regulate the amount of protein which a gene specifies to be made).

Translocation – A translocation occurs when parts of two different chromosomes get pasted together.

Tumor – A synonym for cancer.

Tumor suppressor gene – A gene that specifies (contains the recipe for) a tumor suppressor protein.

Tumor suppressor protein – A protein which, if it ceases to function, can contribute to a cell becoming a cancer cell.

Virus – A parasite which consists of a small amount of genetic information (DNA or RNA) wrapped in a protective coat made of proteins or proteins plus carbohydrates and fats. Viruses can reproduce only by using the machinery of the cells they infect.

Index

Page numbers followed by t indicate a table